Lasers and Their Application in the Cooling and Trapping of Atoms (Second Edition)

Online at: https://doi.org/10.1088/978-0-7503-5482-0

IOP Series in Coherent Sources, Quantum Fundamentals, and Applications

About the Editor

F J Duarte is a laser physicist based in Western New York, USA. His career has covered three continents while contributing within the academic, industrial, and defense sectors. Duarte is editor/author of 15 laser optics books and sole author of three books: *Tunable Laser Optics, Quantum Optics for Engineers*, and *Fundamentals of Quantum Entanglement*. Duarte has made original contributions in the fields of coherent imaging, directed energy, high-power tunable lasers, laser metrology, liquid and solid-state organic gain media, narrow-linewidth tunable laser oscillators, organic semiconductor coherent emission, *N*-slit quantum interferometry, polarization rotation, quantum entanglement, and space-to-space secure interferometric communications. He is also the author of the generalized multiple-prism grating dispersion theory and pioneered the use of Dirac's quantum notation in *N*-slit interferometry and classical optics. His contributions have found applications in numerous fields, including astronomical instrumentation, dispersive optics, femtosecond laser microscopy, geodesics, gravitational lensing, heat transfer, laser isotope separation, laser medicine, laser pulse compression, laser spectroscopy, mathematical transforms, nonlinear optics, polarization optics, and tunable diode-laser design. Duarte was elected Fellow of the Australian Institute of Physics in 1987 and Fellow of the Optical Society of America in 1993. He has received various recognitions, including the *Paul F Foreman Engineering Excellence Award* and the *David Richardson Medal* from the Optical Society.

Coherent Sources, Quantum Fundamentals, and Applications

Since its discovery the laser has found innumerable applications from astronomy to zoology. Subsequently, we have also become familiar with additional sources of coherent radiation such as the free electron laser, optical parametric oscillators, and coherent interferometric emitters. The aim of this book Series in Coherent Sources, Quantum Fundamentals, and Applications is to explore and explain the physics and technology of widely applied sources of coherent radiation and to match them with utilitarian and cutting-edge scientific applications. Coherent sources of interest are those that offer advantages in particular emission characteristics areas such as broad tunability, high spectral coherence, high energy, or high power. An additional area of inclusion are the coherent sources capable of high performance in the miniaturized realm. Understanding of quantum fundamentals can lead to new and better coherent sources and unimagined scientific and technological applications. Application areas of interest include the industrial, commercial, and medical sectors. Also, particular attention is given to scientific applications with a bright future such as coherent spectroscopy, astronomy, biophotonics, space communications, space interferometry, quantum entanglement, and quantum interference.

Publishing benefits
Authors are encouraged to take advantage of the features made possible by electronic publication to enhance the reader experience through the use of color, animation and video, and incorporating supplementary files in their work.

Do you have an idea of a book that you'd like to explore?
For further information and details of submitting book proposals, see iopscience.org/books or contact Ashley Gasque at ashley.gasque@iop.org.

A full list of titles published in this series can be found here: https://iopscience.iop.org/bookListInfo/series-in-coherent-sources-and-applications.

Lasers and Their Application in the Cooling and Trapping of Atoms (Second Edition)

Richard A Dunlap
Dalhousie University, Halifax, Nova Scotia, Canada

IOP Publishing, Bristol, UK

ISBN 978-0-7503-5482-0 (ebook)
ISBN 978-0-7503-5480-6 (print)
ISBN 978-0-7503-5483-7 (myPrint)
ISBN 978-0-7503-5481-3 (mobi)

DOI 10.1088/978-0-7503-5482-0

Version: 20230501

IOP ebooks

British Library Cataloguing-in-Publication Data: A catalogue record for this book is available from the British Library.

Published by IOP Publishing, wholly owned by The Institute of Physics, London

IOP Publishing, No.2 The Distillery, Glassfields, Avon Street, Bristol, BS2 0GR, UK

US Office: IOP Publishing, Inc., 190 North Independence Mall West, Suite 601, Philadelphia, PA 19106, USA

Contents

Preface

Lasers, which were developed in the early 1960s, have not only played an important role in the advancement of scientific knowledge in a wide variety of fields, but have also found commercial applications in a multitude of devices that have become important in our daily lives. These applications include CD/DVD players, laser printers, and fiber optic communication devices. While these devices depend largely on the monochromaticity and coherence of the light which lasers produce, other well-known applications such as laser machining and laser fusion depend on the intensity of laser light. The first part of this book summarizes the physics of lasers and describes some of the more common types of laser and their applications.

Part II of this book considers a somewhat unconventional application of lasers, namely, the cooling and trapping of atoms. The methods by which lasers can be used in conjunction with traditional cooling techniques to achieve ultralow temperatures are discussed. In addition, the combination of lasers with magnetic fields is shown to be an effective means of trapping atoms spatially. These techniques have a number of applications, and some of the most significant are considered. First, the conditions achieved by laser cooling and trapping techniques that are appropriate for the creation of a Bose–Einstein condensate are discussed. This state was first predicted in the 1920s by Satyendra Nath Bose and Albert Einstein. Bose–Einstein condensates were first observed experimentally in 1995 using the techniques presented in this book. Other applications of laser cooling and trapping that are discussed include the manipulation of individual atoms for the purposes of storing data and studying the interactions between atoms in molecular structures. Finally, we consider the use of lasers to form an optical lattice and the means by which atoms trapped in such a lattice can form the basis of a highly accurate time standard.

Author biography

Richard A Dunlap

Richard A Dunlap received a BS in physics from Worcester Polytechnic Institute in 1974, an A.M. in physics from Dartmouth College in 1976, and a PhD in physics from Clark University in 1981. Since receiving his PhD, he has been a faculty member at Dalhousie University. He was appointed Faculty of Science Killam Research Professor in Physics from 2001 to 2006 and served as director of the Dalhousie University Institute for Research in Materials from 2009 to 2015. He currently holds an appointment as research professor in the Department of Physics and Atmospheric Science. Professor Dunlap has published more than 300 refereed research papers, and his research interests have included magnetic materials, amorphous alloys, critical phenomena, hydrogen storage, quasicrystals, superconductivity, and materials for advanced batteries. Much of his work involves the application of nuclear spectroscopic techniques to the investigation of solid-state properties. His previous books include *Experimental Physics: Modern Methods* (Oxford 1988), *The Golden Ratio and Fibonacci Numbers* (World Scientific 1997), *An Introduction to the Physics of Nuclei and Particles* (Brooks/Cole 2004), *Sustainable Energy* (Cengage, 1st edn 2015, 2nd edn 2019), *Novel Microstructures for Solids* (IOP/Morgan & Claypool 2018), *Particle Physics* (IOP/Morgan & Claypool 2018), *The Mössbauer Effect* (IOP/Morgan & Claypool 2019), *Electrons in Solids—Contemporary Topics* (IOP/Morgan & Claypool 2019), *Energy from Nuclear Fusion* (IOP Publishing 2021), and *Transportation Technologies for a Sustainable Future* (IOP Publishing 2023).

Part I

Lasers

Chapter 1

The basic physics of lasers

1.1 Introduction

The term 'laser' is an acronym for '**l**ight **a**mplification by **s**timulated **e**mission of **r**adiation.' The first laser was constructed in 1960 by Thomas Maiman on the basis of theoretical work by Charles Townes and Arthur Schawlow. The operation of the laser follows from the same basic principles as those of its predecessor, the maser (**m**icrowave **a**mplification by **s**timulated **e**mission of **r**adiation). The name 'laser' was originally used to describe devices that worked in the visible portion of the electromagnetic spectrum. However, the term 'light' in its name is now used in a broader sense to indicate any frequency of electromagnetic radiation. Hence, devices which work in the non-visible part of the spectrum are sometimes called 'x-ray lasers,' 'ultraviolet lasers,' 'infrared lasers,' etc. as appropriate, although the term 'maser' is still used for devices that operate in the microwave region.

Lasers are distinguished from other sources of light by the fact that the light that they emit is coherent; in other words, the electromagnetic fields associated with the various photons in the beam are in phase. More precisely, they have a long coherence length, meaning that the beam remains coherent for a large spatial distance. The fact that the light from a laser is coherent means that it can be very intense and directional, and it also requires that the light be polarized and monochromatic. While 'monochromatic' would imply a single frequency or wavelength, it is necessary to consider the extent to which light needs to be monochromatic in order to satisfy the coherence condition. We begin by considering the ways in which visible light can be produced and how the production method affects the degree of monochromaticity.

1.2 Optical spectra

Optical photons are most commonly emitted by materials as a result of electronic processes in the material and can be classified in terms of those which produce a

doi:10.1088/978-0-7503-5482-0ch1

broad distribution of wavelengths and those which are more or less monochromatic. Analogous to the situation in which x-rays are produced by bremsstrahlung, thermal energy results in the emission of photons with a broad distribution of wavelengths, or a continuum spectrum. This is known as blackbody radiation, and the wavelength of the maximum in the energy spectrum is given in nm by Wein's displacement law as follows:

$$\lambda_{\max} = \frac{b}{T},$$

where T is the temperature in Kelvin and b is the displacement constant 2.8978×10^6 K nm. An incandescent lamp produces light that is a good approximation of blackbody radiation, as illustrated in figure 1.1, in which the maximum in the spectrum at about 660 nm corresponds to a temperature of approximately 4400 K. As the visible portion of the electromagnetic spectrum covers the range of wavelengths from about 390 to 700 nm, the figure shows that the incandescent lamp emits most of its radiation in the infrared portion of the spectrum.

Analogous to the discrete line spectra of x-ray sources, which are the result of specific electronic transitions, well-defined spectral lines in the optical region can also result from electronic transitions. Because optical photons have much less energy than x-ray photons, their electronic transitions involve outer-shell electrons with small binding energies, as opposed to the inner-shell electrons involved in x-ray production. The common fluorescent lamp produces spectra which contain well-defined peaks, as illustrated in figure 1.1.

A quantitative example of optical line spectra follows from the calculated energy levels of a hydrogen atom, as illustrated in figure 1.2. The transitions are identified by the energy level of the final (lowest) level involved in the transition and are named

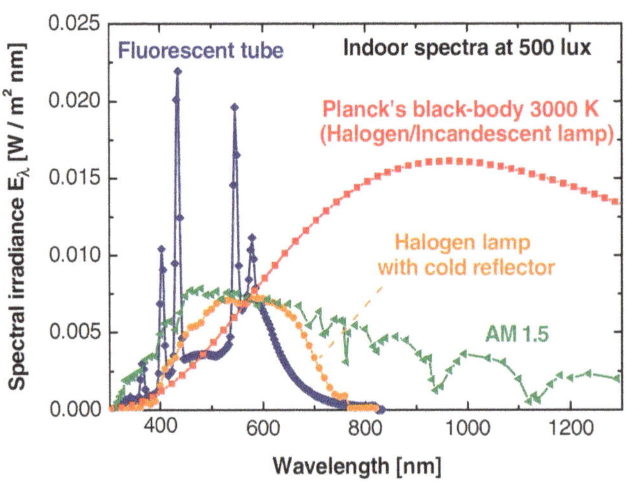

Figure 1.1. Spectra of incandescent, halogen, and fluorescent lamps. AM 1.5 is the simulated spectrum of natural daylight. Reproduced from Virtuani *et al* (2006). Copyright 2006 by Elsevier. Used with permission.

after researchers who studied optical spectroscopy in the early 20th century. The energy, E, of the photon that is emitted is given by

$$E = \frac{m_e e^4}{32\pi^2 \varepsilon_0^2 \hbar^2}\left[\frac{1}{n_2^2} - \frac{1}{n_1^2}\right],$$

where n_2 is the quantum number of the initial (higher-energy) state and n_1 is the quantum number of the final (lower-energy) state. The wavelength of the emitted photons is related to their frequency, f, and the energy of the transition by

$$\lambda = \frac{c}{f} = \frac{hc}{E}. \tag{1.1}$$

When the energy is expressed in eV and the wavelength is expressed in nm, then the constant hc is 1.24×10^3 eV nm. The range of visible wavelengths corresponds to a range of energies from 3.18 to 1.77 eV. The Lyman series involves transitions that end at the ground state ($n_1 = 1$) and produce photons in the far ultraviolet. It is some of the Balmer series transitions (which end at $n_1 = 2$) that have energies corresponding to optical photons.

A typical discharge tube used to produce emission spectra from a gas is illustrated in figure 1.3. A high-voltage discharge excites electrons from their ground state to excited levels. These excited levels are unstable and spontaneously decay back to a lower-energy excited state or the ground state, thereby emitting energy in the form of photons.

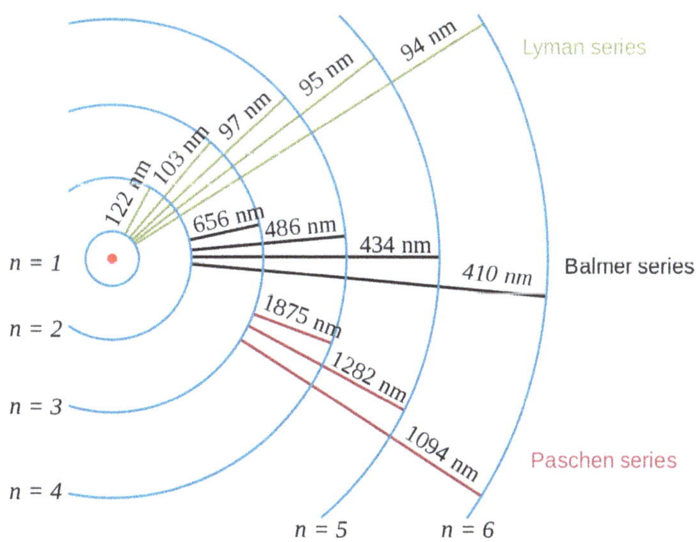

Figure 1.2. The electronic energy levels in atomic hydrogen, showing some of the transitions that give rise to photon emission. The names of the various transition series are indicated. This [Hydrogen transitions] image has been obtained by the author from the Wikimedia website where it was made available under a CC BY 2.0 licence. It is included within this book on that basis. It is attributed to OrangeDog.

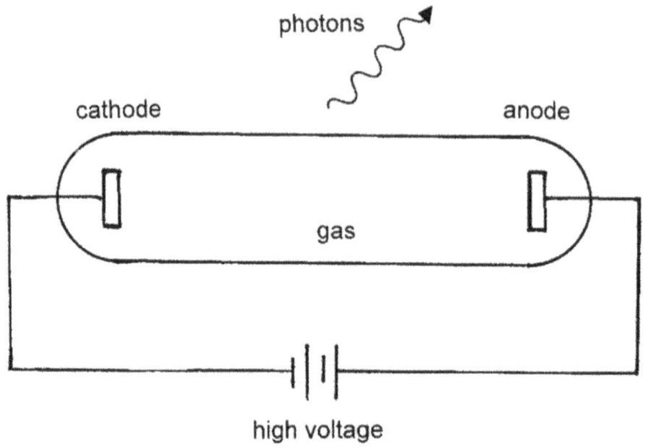

Figure 1.3. A typical gas discharge tube used to produce line spectra.

In cases in which an atom is part of a molecule, then the details of the electronic energy levels become more complex. The energy of the system is not only a result of the energy of the electrons in their quantized energy levels but also results from the rotational and vibrational energy associated with the molecule. These rotational and vibrational levels are quantized; the rotational levels are assigned a quantum number J and the vibrational levels are assigned a quantum number v. Figure 1.4 shows possible transitions involving rotational and vibrational energy levels. Transitions that involve a change in vibrational state, but do not involve a change in rotational state (i.e. $\Delta J = 0$), are referred to as 'Q' transitions. Transitions that involve a change of vibrational state as well as a one-unit decrease in rotational energy (i.e. $\Delta J = -1$) are referred to as 'P' transitions, while those which involve a one-unit increase in rotational energy (i.e. $\Delta J = +1$) are called 'R' transitions. These are identified in figure 1.4.

Figure 1.5 shows the measured spectrum for N_2O and illustrates the P-, Q-, and R-branches of the spectrum. It is common to label the horizontal axis of such spectra with the wave number (usually in cm^{-1}). As the figure shows, the spectral peaks occur for wave numbers at around 600 cm^{-1}. The wave number in cm^{-1} may be expressed in units of energy using the relation that a photon of wave number 1 cm^{-1} has the energy of a photon with a wavelength of 1 cm. This relationship is given by equation (1.1) when the constant is expressed in appropriate units, i.e. $hc = 1.24 \times 10^{-4}$ eV cm. Thus, a wave number of 1 cm^{-1} corresponds to an energy of 1.24×10^{-4} eV and a wave number of 600 cm^{-1} corresponds to an energy of $(1.24 \times 10^{-4}$ eV cm) $\times$ (600 cm^{-1}) = 0.074 eV. The wavelength in cm is merely the inverse of the wave number in cm^{-1}, thus a wave number of 600 cm^{-1} gives a wavelength of (600 cm^{-1}) = 1.67×10^{-3} cm = 1.67×10^{4} nm.

A comparison of figures 1.4 and 1.5 shows that the P-branch has the lowest energy; it therefore appears on the left-hand side of the spectrum. The Q-branch gives rise to the peak at the center of the spectrum. The R-branch corresponds to the right-hand portion of the spectrum.

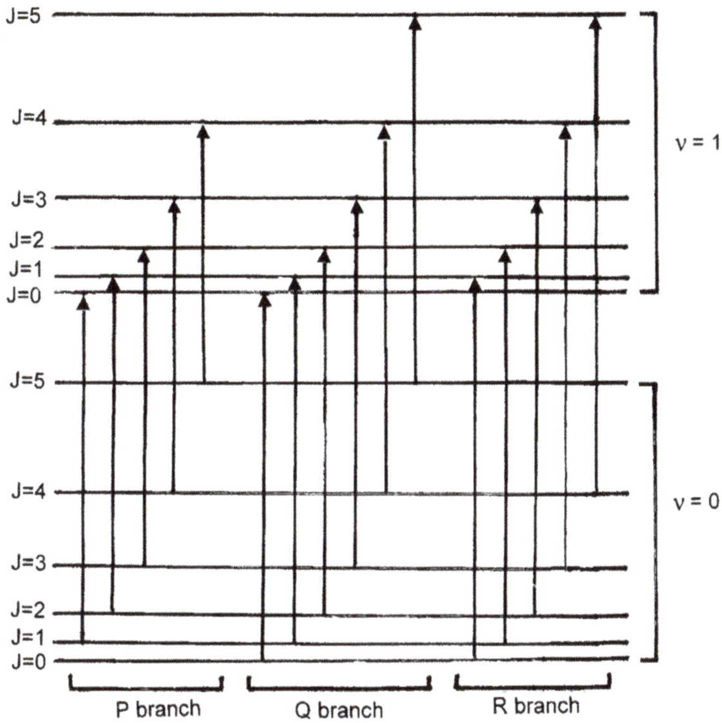

Figure 1.4. Transitions involving vibrational states (designated by the quantum number v) and rotational states (designated by the quantum number J).

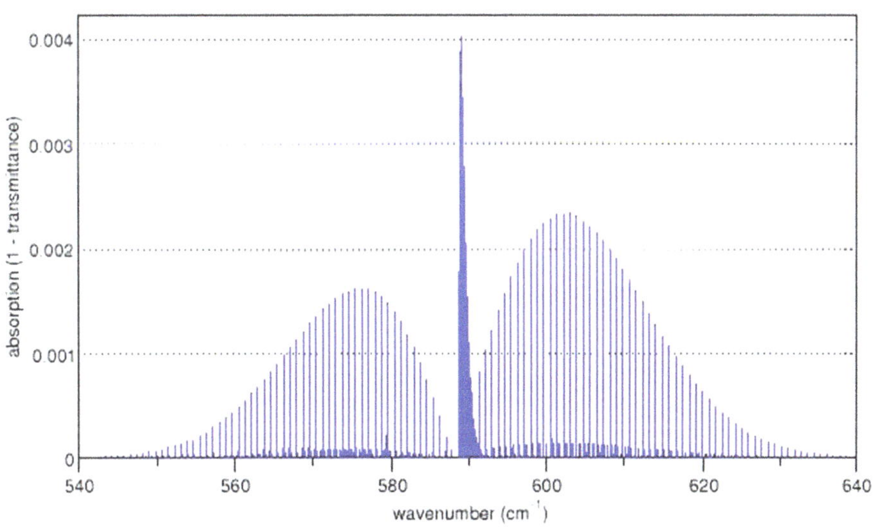

Figure 1.5. The rotational/vibrational spectrum of N_2O. This [Simulated vibration-rotation spectrum] image has been obtained by the author from the Wikimedia website where it was made available under a CC BY-SA 3.0 licence. It is included within this book on that basis. It is attributed to Petergans.

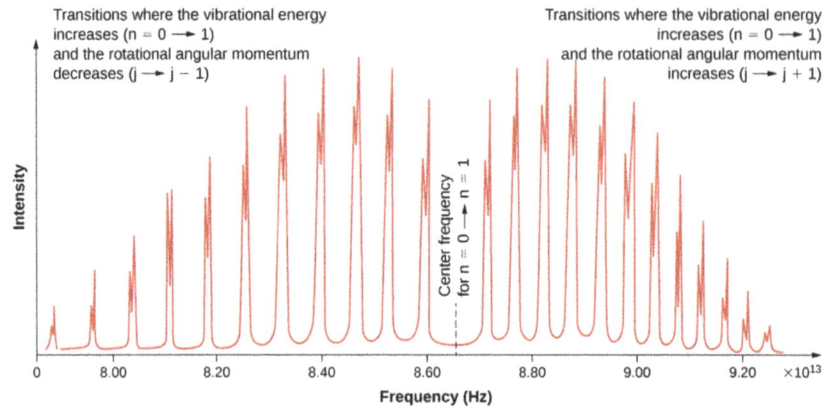

Figure 1.6. The rotational/vibrational spectrum of HCl. Reproduced from [Openstax]. CC BY 4.0.

The rotational/vibrational spectrum of N_2O. spectrum illustrates an interesting feature that is often seen in such spectra. The P- and R-branches show the presence of a weak spectral component superimposed on the principal spectrum. Nitrogen consists of two naturally occurring isotopes, ^{14}N (about 99.7%) and ^{15}N (about 0.3%). These two isotopes have different masses and give rise to molecules with different moments of inertia. Thus, the splitting of the energy levels is slightly different for molecules which contain atoms of different nitrogen isotopes, and this results in two superimposed spectra.

Figure 1.6 shows the rotational/vibrational spectrum of HCl. It is clear in this figure that the P- and R-branches appear, as expected, but the Q-branch is missing. For a linear molecule consisting of two elements, A and B, of composition AB, quantum mechanical selection rules forbid transitions with $\Delta J = 0$, thereby eliminating the Q-branch.

1.3 Stimulated emission

The previous section has shown how electronic transitions in atoms and molecules that involve discrete quantized energy levels can give rise to well-defined spectral wavelengths. While this feature more or less satisfies the requirement that a light source should produce photons that are monochromatic in order to be classified as a laser, we still have to consider the nature of the electronic transition and the question of polarization. Here, we begin with a consideration of the difference between spontaneous emission processes and stimulated emission processes.

Figure 1.7 shows a diagram of a spontaneous emission process. In figure 1.7(a), energy (in this case in the form of a photon) is absorbed by an atom, giving rise to the excitation of one of its electrons. As this excited state is unstable, it decays back to the ground state as shown in figure 1.7(b) after some time, re-emitting a photon with an energy equal to the difference between the energies of the excited and ground states.

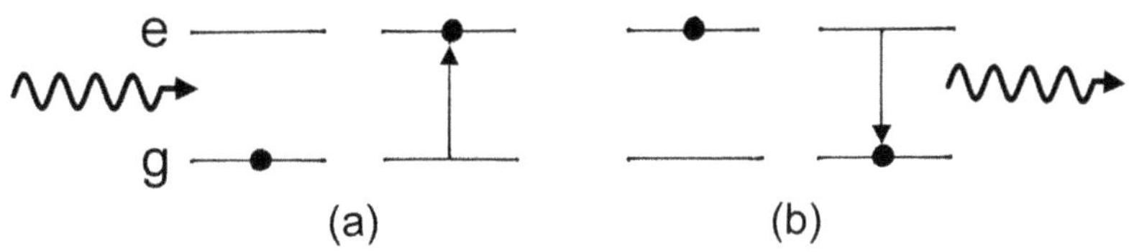

Figure 1.7. A spontaneous emission process: (a) the excitation of an atom into an excited state and (b) spontaneous decay back to the ground state accompanied by photon emission.

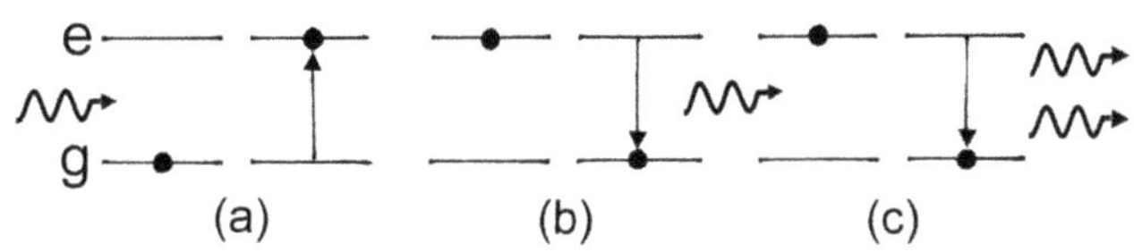

Figure 1.8. A stimulated emission process: (a) the absorption of a photon and the excitation of an electron into an excited state, (b) spontaneous decay from an unstable excited state and the emission of a photon, and (c) the stimulated emission of a photon from another atom in an excited state by the photon produced by spontaneous decay in panel (b).

Figure 1.8 illustrates a stimulated emission process. As in the spontaneous emission process, an atom absorbs energy, it is excited, and this excited state decays spontaneously, re-emitting a photon. This re-emitted photon encounters another atom of the same type which is in the excited state, and which has not yet decayed spontaneously. The incident photon causes the excited atom to decay before it would have decayed spontaneously. As the two atoms involved are of the same type, they have the same energy levels, and the incident photon and the photon that results from the stimulated emission have the same energy (within certain limits, as discussed further below). There are two additional features that are of importance in this process: the two photons not only have the same energy, but they are in phase and have the same polarization vector. Thus, they satisfy the description of radiation from a laser.

The characteristics of the radiation described above are that it is monochromatic (to the degree discussed below) and it is intense, as the components of the beam are all in phase and interfere constructively with each other. If this process continues from a collection of the same type of atom, then the beam continues to increase in intensity. Light in which the components of the beam remain in phase with one another over an extended distance is referred to as coherent.

Two practical problems need to be resolved in order to create the situation described above. First, we need to ensure that there are a sufficient number of atoms in the excited state (rather than in the ground state) for the stimulated emission to occur frequently, and we need to ensure that there is sufficient opportunity for the beam to gain strength by successive stimulated emissions. The solutions to these two problems are the creation of a population inversion and the creation of an appropriate resonant cavity, respectively. These two requirements are discussed below.

1.4 Creating a population inversion

Consider a simple two-level quantum system as illustrated in figure 1.9(a). At zero temperature, all atoms are in their ground state. However, at any temperature, T, above absolute zero, thermal effects need to be considered. The distribution of particle velocities as a function of temperature is given by the Maxwell–Boltzmann distribution, as illustrated in figure 1.10. This distribution has the following functional form:

$$f(v) = \left(\frac{m}{2\pi k_B T}\right)^{3/2} 4\pi v^2 \exp\left(-\frac{mv^2}{2k_B T}\right),$$

where $f(v)$ is the fraction of atoms with velocity v and k_B is the Boltzmann constant. The Maxwell–Boltzmann function may also be expressed as a distribution of energy using a change of variables and the relation $E = mv^2/2$, as follows:

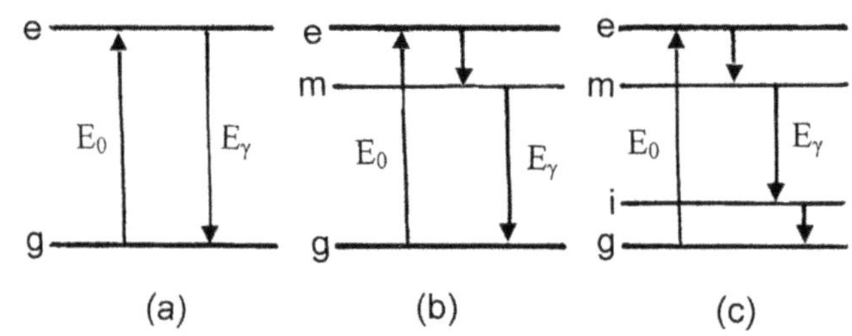

Figure 1.9. (a) A two-level system showing energy absorption, E_0, and emission, E_γ, between the ground state (**g**) and the excited state (**e**). (b) A three-level system showing energy absorption between the ground state (**g**) and the excited state (**e**), spontaneous emission between the excited state (**e**) and a metastable state (**m**), and stimulated emission between the metastable state (**m**) and the ground state (**g**). (c) A four-level system showing absorption between the ground state (**g**) and the excited state (**e**), spontaneous emission between the excited state (**e**) and the metastable state (**m**), stimulated emission between the metastable state (**m**) and the intermediate state (**i**), and spontaneous emission between the intermediate state (**i**) and the ground state (**g**).

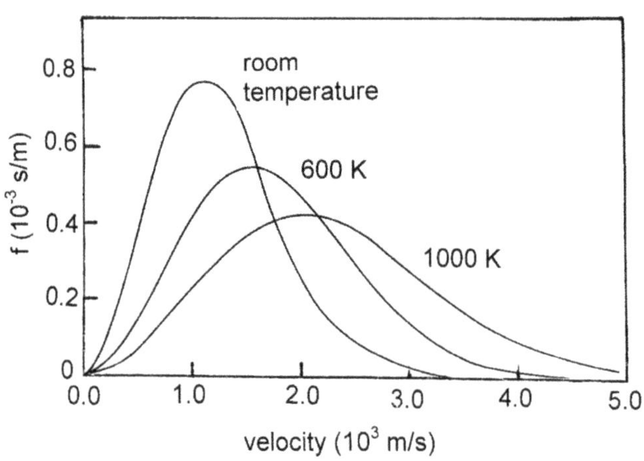

Figure 1.10. Maxwell–Boltzmann distributions at different temperatures.

$$f(E) = 2\sqrt{\frac{E}{\pi}}\left(\frac{1}{k_B T}\right)^{3/2} \exp\left(-\frac{E}{k_B T}\right).$$

For a system consisting of discrete microstates with energies E_i, the number of atoms in a state with energy E_i follows from the Maxwell–Boltzmann distribution

$$\frac{N_i}{N} = \frac{\exp\left(-\frac{E_i}{k_B T}\right)}{\sum_j \exp\left(-\frac{E_j}{k_B T}\right)}, \tag{1.2}$$

where N is the total number of atoms in the system. A plot of equation (1.2) at room temperature is shown in figure 1.11. It is clear that the state's population decreases as a function of increasing energy. This is the normal equilibrium situation. For a population inversion, we require that the population at a particular energy level is greater than that of the level below it. For the case shown in figure 1.9(a), atoms in the ground state can be excited into the higher-energy state by the input of an energy greater than the excited-state energy, E_0. This can be done by (e.g.) photon irradiation or an electric discharge. Excited-state atoms spontaneously decay back to the ground state after a characteristic time (the mean life of the excited state), releasing energy in the form of a photon of energy $E_\gamma = E_0$. If a sufficient number of atoms are excited from the ground state in less time than the mean life of the excited state, population inversion results, and stimulated emission can occur. However, this population inversion cannot be maintained, as stimulated emissions decrease the population of the excited state and increase the population of the ground state until the population inversion is eliminated.

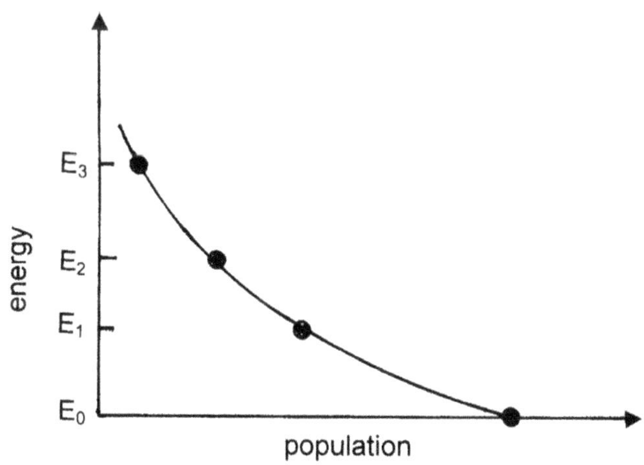

Figure 1.11. Population as a function of energy for the Maxwell–Boltzmann distribution near room temperature.

The simplest system which is compatible with the requirements for the design of an operational laser is the three-level system, as shown in figure 1.11(b). The short-lived (e.g., ~10^{-8} s) excited state decays spontaneously to a metastable state at a lower energy, which in turn decays by stimulated emission to the ground state. The metastable state is relatively long-lived (e.g., 10^{-5}–10^{-4} s). Thus, ground-state atoms are pumped up to the excited state which quickly decays to the metastable state, where atoms collect to create a population inversion relative to the ground state. Three-level lasers most commonly operate in a pulsed mode: a short discharge from a flash tube, for example, pumps the system up to the excited state from the ground state, which is followed by stimulated emission from the metastable state. It should be noted that the energy (per photon) required to excite the ground-state atoms up into the excited state is greater than the energy (per photon) produced by the stimulated decay. This means, for example, that green photons might be needed to excite the transition to the excited state, while red photons might be emitted during the stimulated decay process.

Lasers that operate continuously are more commonly based on a four-level system, as shown in figure 1.9(c). In this case, atoms that are pumped up to the excited state decay quickly by spontaneous emission to the relatively long-lived metastable state. Atoms collect in this metastable state and then decay by stimulated emission to a short-lived intermediate state, which quickly decays back to the ground state. Since the intermediate state decays quickly, its population remains small. The population inversion can, therefore, be maintained between the metastable state and the intermediate state and the excited state can continuously be populated by the excitation of ground-state atoms.

1.5 Laser modes and coherence

The second consideration in laser design is the creation of a resonant cavity so that the beam can be amplified by continuous constructive interference of the beam components. Figure 1.12 shows the basic design of a resonant cavity. The cavity has a fully reflecting mirror at one end and a partially reflecting mirror at the other end. Photons that are emitted as a result of the stimulated decay process are reflected back and forth in the cavity, causing more stimulated decays, which produce more photons, all of which are of the same energy and are in phase with one another. Some of the coherent radiation leaks out of the partially reflecting mirror, thus forming the laser beam.

The above process sounds quite simple, but there is a serious concern; how do we maintain the phase relationship of the waves that are reflected from the ends of the cavity? In order to do this, it is necessary to produce a resonant cavity that is a half-integer number of wavelengths in length. In other words, we want to create conditions in which the waves inside the cavity are standing waves. Figure 1.12 illustrates this point. For this condition to be met, we require that

$$n\frac{\lambda}{2} = L \tag{1.3}$$

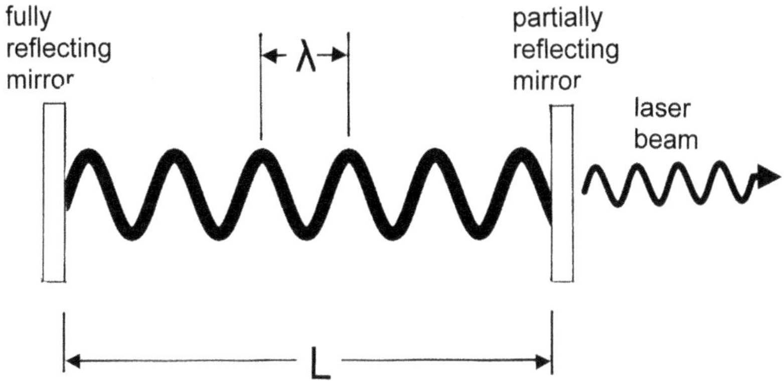

Figure 1.12. The conditions used to create a resonant laser cavity which is a half-integer number of wavelengths long.

where n is an integer, λ is the wavelength of the radiation, and L is the length of the cavity. In fact, figure 1.12 is not really to scale, as a typical wavelength would be in the range of a few hundred nanometers and a typical cavity have a length in the centimeter or many-centimeter range. Using $L = 30$ cm and $\lambda = 500$ nm gives $(3 \times 10^{-1}\ \text{m})/(2.5 \times 10^{-7}\ \text{m}) = 1.2 \times 10^{6}$ half wavelengths in the cavity. If we think about how we might actually create this situation, we realize that it is a very difficult task. The light, which has a wavelength of the order of a few hundred nanometers, bounces back and forth in the cavity many times. Any error in satisfying the relationship in equation (1.3) will be compounded with each reflection. Therefore, we, might think that we need to know how accurately equation (1.3) has to be satisfied in order to make a functioning laser. Fortunately, we never actually have to concern ourselves with this problem, which brings us back to a question we have never really answered: how monochromatic is monochromatic?

The decay of a quantum mechanical state is a statistical process characterized by a characteristic mean decay time, τ. As the exact time of the decay is uncertain, the energy of the state is uncertain with a distribution that is related to the mean time by the Heisenberg uncertainty principle:

$$\Delta E \Delta t = \Delta E \tau \geqslant \frac{\hbar}{2}$$

or

$$\Delta E \geqslant \frac{\hbar}{2\tau} \tag{1.4}$$

This form of the Heisenberg uncertainty principle follows directly from the more common form involving position, x, and momentum, p:

$$\Delta x \Delta p \geqslant \frac{\hbar}{2}, \tag{1.5}$$

by the relationships for position, velocity, v, and time, t:

$$x = vt, \tag{1.6}$$

and momentum and energy, E:

$$p = \sqrt{2mE}, \tag{1.7}$$

where m is mass. Taking the differentials of equations (1.6) and (1.7) and substituting them into equation (1.5) gives the form of the uncertainty relation shown in equation (1.4). Using $\tau = 10^{-5}$ s for a metastable-state mean lifetime in equation (1.4), we find $\Delta E \geqslant 3 \times 10^{-11}$ eV, out of a photon energy of about 2.5 eV (for $\lambda = 500$ nm). This value is actually a substantial underestimate of the actual width of the energy distribution, as thermal effects substantially broaden the distribution. In a gas, this broadening results from the Doppler effect due to thermal motion. In a solid, phonon (that is quantized lattice vibrations) interactions broaden the energy distribution. In many materials, an energy width of around 10^{-5} eV is fairly typical. Since

$$E = \frac{hc}{\lambda},$$

it is straightforward to show that

$$\Delta\lambda = -hc\frac{\Delta E}{E^2}.$$

From the above example, we find that $\Delta\lambda \sim 2 \times 10^{-3}$ nm. This is still small compared to the wavelength of 500 nm, so the photons are monoenergetic to about one part in $(500 \text{ nm})/(2 \times 10^{-3} \text{ nm}) = 2.5 \times 10^5$.

It is now necessary to consider the resonant modes of the cavity. We need to ask: if the cavity contains n half wavelengths, then how much does the wavelength have to change to fit n+1 half wavelengths in the cavity? From equation (1.3), we solve for λ:

$$\lambda = \frac{2L}{n}$$

and differentiating gives

$$\Delta\lambda = -\frac{2L}{n^2}\Delta n = -\frac{\lambda^2}{2L}\Delta n. \tag{1.8}$$

So, for $\Delta n = 1$, then $\Delta\lambda = 4 \times 10^{-4}$ nm. Comparing this with the typical width of the wavelength distribution from the stimulated emission shows that there are about five cavity modes within this distribution. This relationship is illustrated in figure 1.13.

The above analysis shows that in a typical situation, it is unnecessary to adjust the length of the resonant cavity to match the wavelength of the lasing transition. No matter what the length of the cavity is, it contains several resonant modes. A measured wavelength spectrum of a laser transition is illustrated in figure 1.14. Thus,

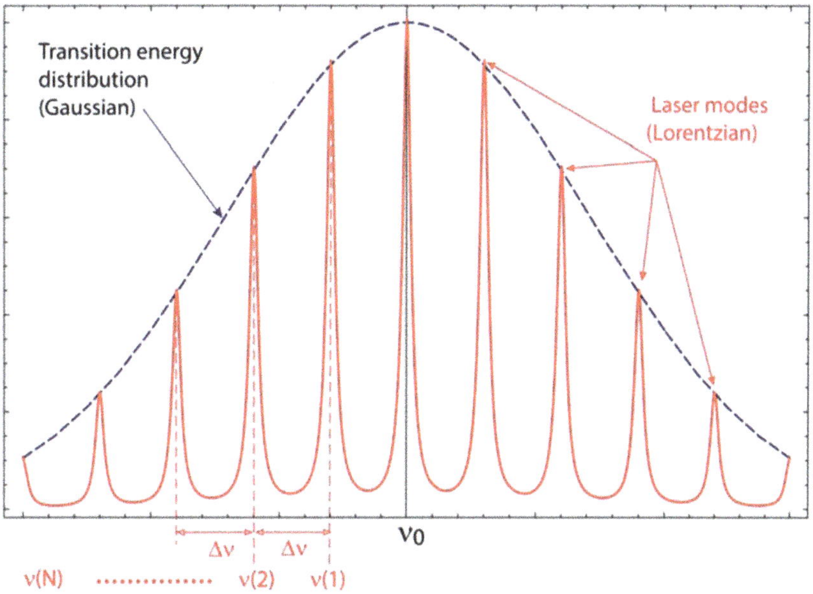

Figure 1.13. The relationship between the wavelength distribution of stimulated emission and resonant cavity modes in a typical laser. This [Schematic representation of longitudinal laser modes] image has been obtained by the author from the Wikimedia website where it was made available under a CC BY-SA 3.0 licence. It is included within this book on that basis. It is attributed to Dr. Wolfgang Geithner. (Labels translated to English).

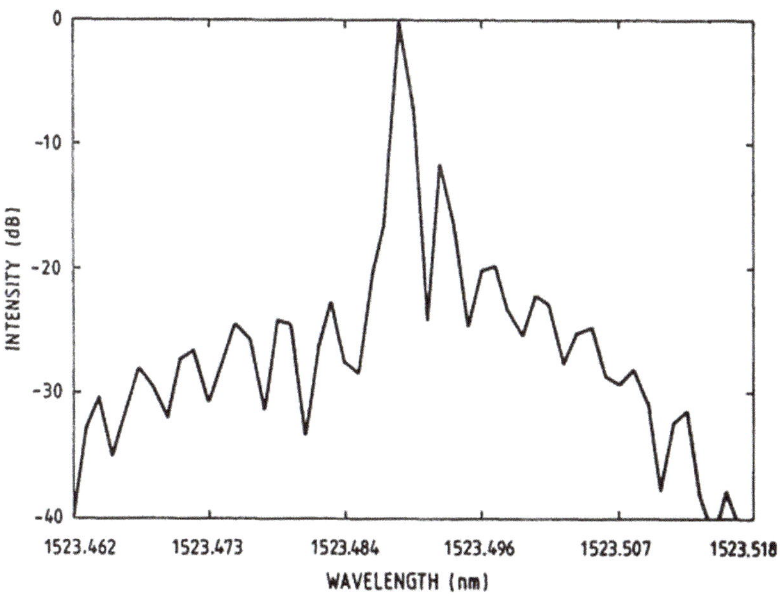

Figure 1.14. Wavelength spectrum of the lasing transition at 1523 nm (infrared) of a He–Ne laser, showing the presence of cavity modes. Reproduced from Junttila and Stahlberg (1990).

the monochromatic light produced by a laser typically consists of a number of closely spaced modes within a Gaussian-like envelope. In the next two chapters, we describe the operational principles of several common types of laser.

References and suggestions for further reading

Csele M 2004 *Fundamentals of Light Sources and Lasers* (Hoboken, NJ: Wiley)

Geithner W 2006 Schematic representation of longitudinal laser modes https://de.wikipedia.org/wiki/Datei:LaserModes.jpg

Hecht J 1992 *The Laser Guidebook* 2nd edn (New York: McGraw-Hill)

Junttila M L and Stahlberg B 1990 Laser wavelength measurement with a Fourier transform wavemeter *Appl. Opt.* **29** 3510–6

Openstax 2022 Molecular Spectra https://phys.libretexts.org/@go/page/4542

OrangeDog 2009 Hydrogen transitions https://commons.wikimedia.org/wiki/File:Hydrogen_transitions.svg

Petergans 2013 Simulated vibration-rotation spectrum https://commons.wikimedia.org/wiki/File:Nu2_nitrous_oxide.png

Silfvast W T 1996 *Laser Fundamentals* (Cambridge: Cambridge University Press)

Svelto O 1998 *Principles of Lasers* 4th edn (New York: Springer)

Virtuani A, Lotter E and Powalla M 2006 Influence of the light source on the low-irradiance performance of Cu(In,Ga)Se_2 solar cells *Sol. Energy Mater. Sol. Cells* **90** 2141–9

Wilson J and Hawkes J F B 1987 *Lasers: Principles and Applications* (New York: Prentice-Hall)

IOP Publishing

Lasers and Their Application in the Cooling and Trapping of Atoms (Second Edition)

Richard A Dunlap

Chapter 2

Conventional lasers

2.1 Introduction

The earliest lasers used lasing atoms contained in a solid material. The most notable of these was the ruby laser. Later, traditional lasing transitions in gaseous and liquid media were utilized in the construction of lasers. This chapter reviews the most important of these lasers.

2.2 Solid-state lasers

Solid-state lasers contain atoms that undergo a lasing transition that are contained within a solid matrix. The earliest lasers that were developed were solid-state lasers; specifically, the ruby laser. Ruby is aluminum oxide (Al_2O_3), which is itself colorless. Cr^{3+} ion impurities in aluminum oxide produce the characteristic red color of ruby. It is the Cr^{3+} ion which has energy levels that produce the laser radiation. Figure 2.1 shows the energy levels of the Cr^{3+} ion. There are two short-lived states that readily absorb photons at around 400 and 550 nm (corresponding to blue and green light, respectively). These short-lived states decay to a metastable state which has a lifetime of about 3 ms. This metastable state decays back to the ground state and emits photons at 694 nm (red). In this sequence, Cr^{3+} ions are pumped from the ground state to the two short-lived states, which then populate the metastable state, creating a population inversion between the metastable state and the ground state. This population inversion results in stimulated emission, returning the ions to the ground state. Since the stimulated emission repopulates the ground state, the population inversion cannot be continuously maintained, and the ruby laser must be operated in a pulsed mode.

Solid-state lasers are typically pumped using photons from a xenon flash tube. These flash tubes are intense, and a substantial portion of their spectrum falls within the proper wavelength range, as illustrated in figure 2.2. The figure shows that there

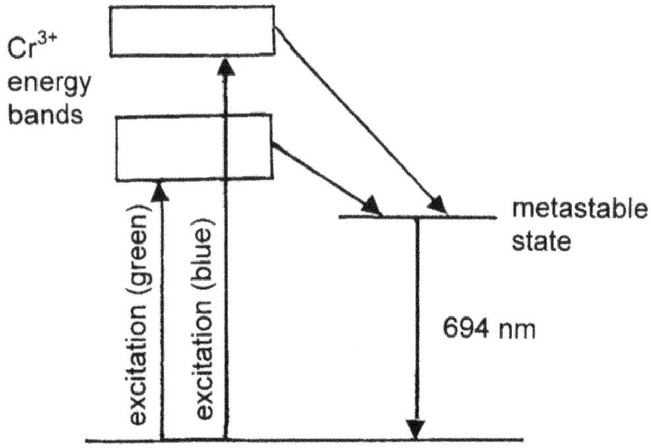

Figure 2.1. A simplified energy level diagram of the Cr^{3+} ion, showing the stimulated emission at a wavelength of 694 nm.

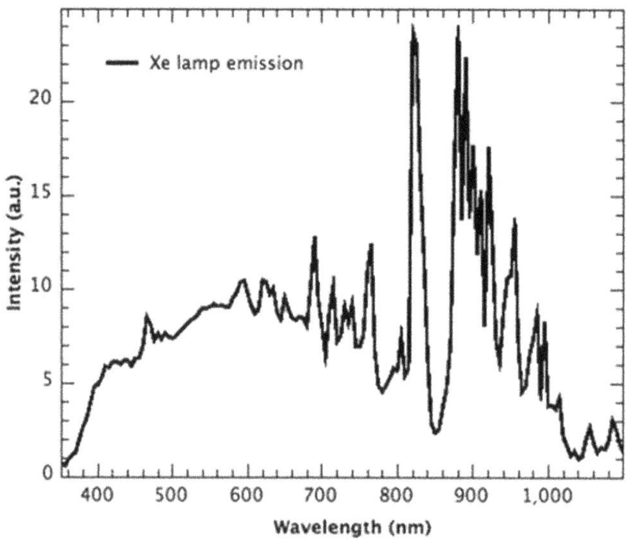

Figure 2.2. The spectral output of a xenon flash tube. Reproduced from Presciutti *et al* (2014). CC BY-SA 4.0.

are some discrete peaks in the long-wavelength infrared portion of the spectrum. In the near-infrared, visible, and near-ultraviolet portions of the spectrum (from about 800 nm down to about 300 nm) there is a broad, mostly featureless continuum. The spectrum is quite intense in the 400–550 nm region required to pump the ruby laser.

One of the most common solid-state lasers is the Nd^{3+}:YAG laser. In this laser, Nd^{3+} ions are responsible for the lasing transitions. These are incorporated into a suitable transparent matrix. Yttrium aluminum garnet (YAG), $Y_3Al_5O_{12}$, is commonly used, but other materials such as $LiYF_4$, YVO_4, and glass have also been used. An energy level diagram of the Nd^{3+} ion is shown in figure 2.3. This ion

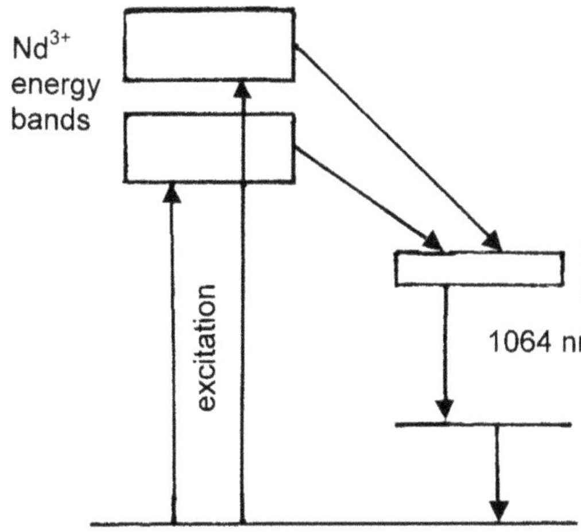

Figure 2.3. The energy levels of Nd^{3+} in YAG.

has four energy levels and can, in principle, be used either in the pulsed or the continuous mode of operation. Photons from a flash lamp in the 730–800 nm range (the near-infrared) are used to pump the Nd^{3+} ions from the ground state up to several excited states, as shown in the figure. These states spontaneously decay to the metastable state, which decays by stimulated emission to the short-lived intermediate state. The metastable and intermediate states have some splitting, so there are several closely spaced lasing transitions at around 1064 nm.

Nd:YAG lasers have found a number of applications as discussed further below, because quite powerful lasers of this type can be constructed fairly economically. However, the wavelength produced by Nd:YAG lasers (and all Nd^{3+}-based lasers) is in the infrared. For some applications, this is inconvenient, as photons in the visible portion of the spectrum are more appropriate. One way of dealing with this difficulty is to use harmonic generation, as discussed below.

2.3 Harmonic generation

The photons produced by a Nd:YAG laser at 1064 nm have an energy of 1.17 eV. These may be converted into optical photons by second-harmonic generation or frequency doubling. This process is easily explained from a quantum mechanical point of view. Two photons of 1.17 eV (a wavelength of 1064 nm) can be combined to produce one photon at 2.34 eV (a wavelength of 532 nm). The number of photons is reduced by a factor of two, but the energy per photon is doubled.

This process can be viewed from a more classical standpoint by looking at the effects of the electric field associated with an electromagnetic wave as it passes through a dielectric material. In a dielectric material, the positive and negative charge distributions are not coincident, and this gives rise to an electric dipole moment. Under normal conditions, this results in a polarization, P, which is proportional to the electric field, E,

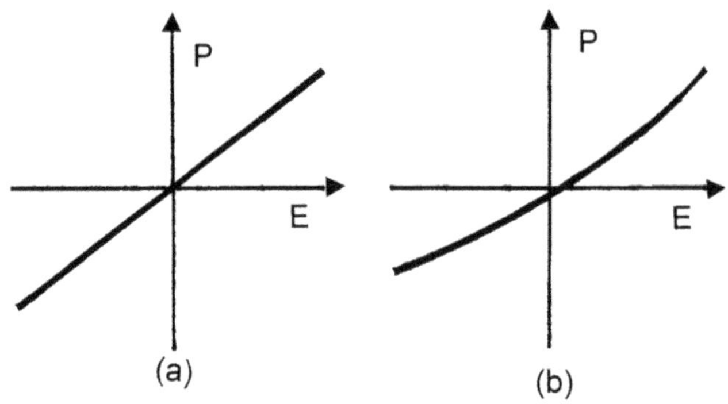

Figure 2.4. The relationship between the electric field associated with a beam of light and the induced polarization for (a) a linear medium and (b) a nonlinear medium.

$$P = \chi_1 \varepsilon_0 E, \tag{2.1}$$

where χ is the relative electric susceptibility of the material and ε_0 is the permittivity of free space. In this context, the polarization is the vector sum of the induced dipole moments in the material and is not the same as the polarization of the light beam (which represents the orientation of the electric field vectors). The relationship in equation (2.1) shows a linear dependence between P and E, as illustrated in figure 2.4(a). If the electric field associated with the light oscillates sinusoidally, then the induced polarization also oscillates sinusoidally (at the same frequency), generating a sinusoidally oscillating electric field that contributes to the propagating electric field associated with the light beam.

In some materials, and particularly for large electric fields, the relationship between the electric field and the polarization is nonlinear, as shown in figure 2.4(b). The nonlinear behavior of the polarization can be expressed as a Taylor expansion in powers of the electric field as follows:

$$P = \chi_1 \varepsilon_0 E + \chi_2 \varepsilon_0 E^2 + \chi_3 \varepsilon_0 E^3 + \cdots. \tag{2.2}$$

This means that if the electric field varies sinusoidally, the induced polarization is not a pure sine wave. Thus, the electric field associated with the polarization is also not purely sinusoidal. This non-sinusoidal electric field can be expanded in a Fourier series of multiples of the fundamental frequency, f_0, of the electric field. Figure 2.5(a) shows the Fourier expansion of a sinusoidal electric field with a single Fourier component at the fundamental frequency. Figure 2.5(b) shows the Fourier expansion of a non-sinusoidal electric field.

The harmonics of the fundamental frequency represent components of the beam that have twice, three times, etc. the frequency (and energy) of the incident beam. Some materials are quite efficient at generating harmonics because of a significant nonlinear component that results from the nonlinear terms for the polarization in equation (2.2). The harmonic at twice the fundamental frequency is generally the most intense harmonic and corresponds to second-harmonic generation. This situation is illustrated in figure 2.6, in which light at the fundamental frequency is

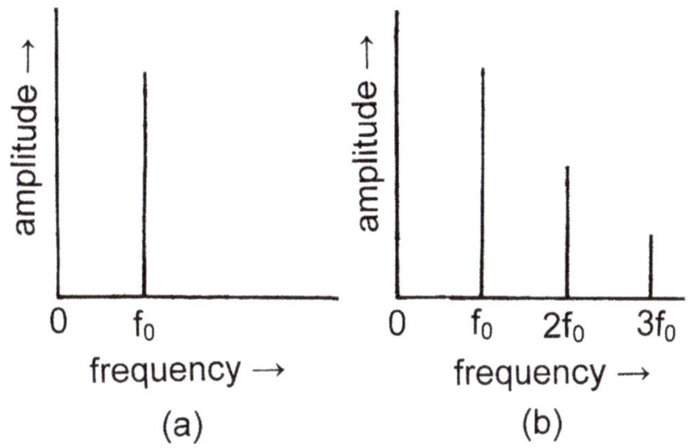

Figure 2.5. Fourier expansion of (a) a linear polarization and (b) a nonlinear polarization.

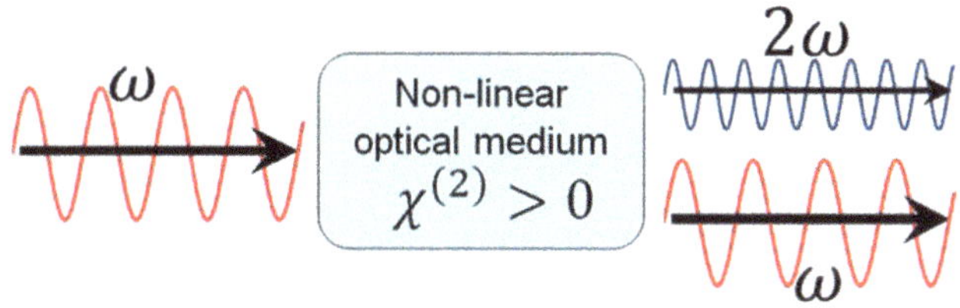

Figure 2.6. Second-harmonic generation showing input and output frequencies. CC BY-SA 4.0. This [Schematic view of the SHG conversion of an exciting wave in a nonlinear medium with a non-zero second-order nonlinear susceptibility] image has been obtained by the author from the Wikimedia website where it was made available under a CC BY-SA 4.0 licence. It is included within this book on that basis. It is attributed to BP-Aegirsson.

input into a nonlinear optical material and the output consists primarily of a mixture of light at the fundamental frequency and light at twice the fundamental frequency.

The efficiency of second-harmonic generation may be viewed in a quantum mechanical sense as the fraction of photons that combine to form photons of twice the energy. In the classical sense, the second-harmonic generation efficiency may be viewed as the ratio of the $2f_0$ beam component to the initial beam intensity. An example of second-harmonic generation efficiency for the conversion of 1064 nm near-infrared radiation to 532 nm green light is shown in figure 2.7. Up to some point, the polarization becomes progressively more nonlinear as a function of the electric field and the second-harmonic generation efficiency increases.

In some instances, it is convenient or necessary to obtain light at a frequency that is more than the twice the output frequency of the laser. In such cases, frequency tripling (or third-harmonic generation) can be a viable option. Frequency tripling is a method of combining the fundamental and second-harmonic outputs from a nonlinear medium as shown in figure 2.6 to produce a laser beam at three times the

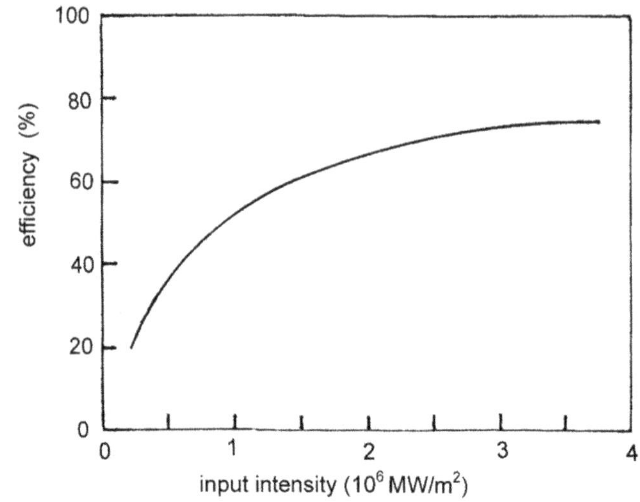

Figure 2.7. Second-harmonic generation efficiency for three layers of magnesium oxide:lithium tantalate crystals for 1064 nm Nd:YAG laser radiation.

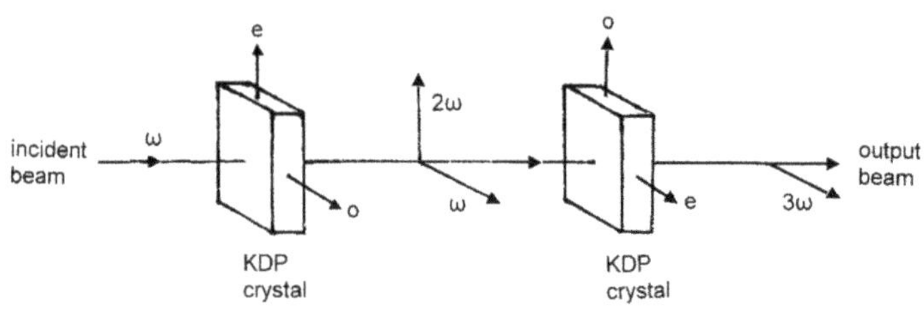

Figure 2.8. Frequency tripling. The ordinary (o) and extraordinary (e) directions of the KDP crystals are shown. Reprinted from Dunlap (2021) © IOP Publishing. Reproduced with permission. All rights reserved.

fundamental frequency. Quantum mechanically, this corresponds to combining the fundamental frequency photons and the second-harmonic photons to yield photons at three times the fundamental photon energy. An example of this approach is the frequency tripling of the near-infrared 1064 nm radiation from a Nd:YAG laser to produce near-ultraviolet radiation at around 354 nm. This approach can have efficiencies of around 50%. Frequency tripling is used for inertial confinement fusion experiments, as the near-ultraviolet radiation is more readily absorbed by the fusion fuel than near-infrared radiation. This application of frequency tripling is discussed further in section 4.7. A practical system for frequency tripling utilizes two nonlinear optical crystals as shown in figure 2.8. The first crystal produces beam components at the fundamental frequency and the second-harmonic frequency, while the second crystal combines these to produce the third-harmonic beam. The most common nonlinear material used for such systems is monopotassium phosphate (KH_2PO_4), also known as potassium dihydrogen phosphate (KDP).

2.4 Gas lasers

Gas lasers use lasing media which are in the gaseous phase, often in combination with other gas atoms or molecules. The gas used may consist of either neutral atoms, ions, or molecules. Examples of lasers utilizing neutral atoms and molecules are discussed in this section. Gas lasers are analogous to gas discharge tubes, as shown in figure 1.3, except that they are designed to emit light by stimulated, rather than spontaneous, emission. Excitation may be induced using electric fields, as in figure 1.3, or by the application of radio-frequency radiation.

The helium–neon (He–Ne) laser is the most common of the gas lasers in current use. The lasing cavity contains a mixture of about ten parts He to one part Ne. Although it is actually transitions in the neon atoms that produce the laser radiation, the large amount of helium that is present is necessary in order to create a population inversion. Energy level diagrams of He and Ne are shown in figure 2.9. The energy associated with the electric discharge preferentially excites the helium atoms (because there are more of them) to two excited states, as shown in the figure. These two helium energy levels are at nearly the same energy as two levels in neon. Through atomic collisions, energy is transferred from excited helium atoms (which fall back to their ground state) to ground-state neon atoms (which are pumped up into their excited states). These are the metastable laser states in the neon. The reason that the neon atoms are not directly excited through an electric discharge is that there are short-lived states (the laser intermediate states) that would be populated as well. The transfer of energy from the helium to the neon creates a population inversion between several combinations of neon states, i.e. between the 5s and 4p states, between the 5s and 3p states, and between the 4s and 3p states, as illustrated in the diagram. Stimulated emission can now occur between these combinations of metastable and short-lived intermediate states. Since there is fine splitting of some of these levels, closely spaced lasing transitions sometimes result.

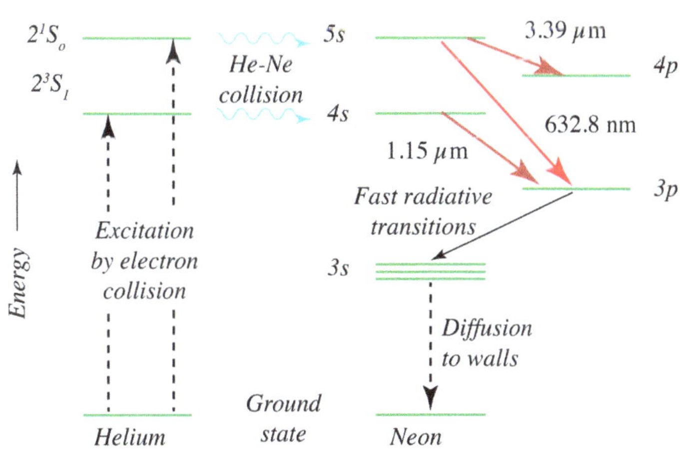

Figure 2.9. Energy levels in the He–Ne laser. CC BY-SA 4.0 by XuPanda. This [Energy levels in a Laser] image has been obtained by the author from the Wikimedia website where it was made available under a CC BY-SA 4.0 licence. It is included within this book on that basis. It is attributed to XuPanda.

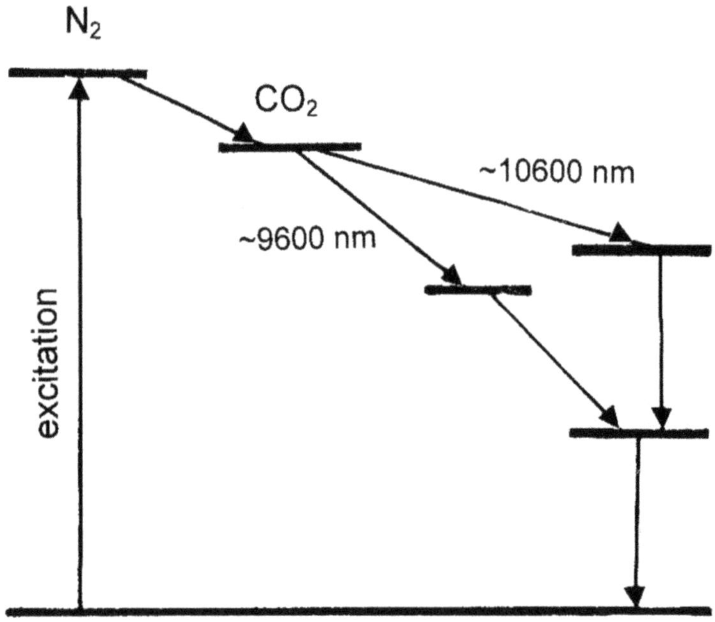

Figure 2.10. The energy levels in a CO_2 laser, showing the role of nitrogen and the laser transitions at 10 600 and 9600 nm.

The most intense and most commonly used lasing transition in neon is the 633 nm (red) transition from the 5s to 3p state. Weaker lines are produced in the green (543 nm) and infrared (1118, 1152, 1523, and 3391 nm) regions.

The carbon dioxide laser is a common example of a molecular laser. This laser contains a mixture of 10%–20% CO_2 and 10%–20% nitrogen. The remainder of the gas is mostly helium with (sometimes) a small amount of hydrogen or xenon. The energy diagram for CO_2 and nitrogen is shown in figure 2.10. The lasing transitions are in the CO_2 states, but the nitrogen is necessary to create the population inversion, in much the same way that helium is needed in the He–Ne laser. Energy from an electric discharge excites nitrogen molecules, N_2, from their ground state to a higher energy level. This energy is transferred to CO_2 molecules through collisions and populates the metastable state. This metastable state decays by stimulated emission to one of several short-lived intermediate states. The helium is needed to help carry away excess energy from the nitrogen molecules after they have interacted with the CO_2.

2.5 Dye lasers

Dye lasers are analogous to molecular sources of optical spectra. They are referred to as dye lasers because the earliest lasers of this type utilized transitions in commercial dyes to produce laser radiation. The dyes and other organic molecules used in this type of laser have very complex electronic energy levels, giving rise to a very large number of possible transitions that can emit stimulated radiation.

Dye lasers generally consist of the 'dye' molecules dissolved in a liquid organic solvent. The pumping radiation can be produced by a broadband flash lamp or a traditional solid-state or gas laser. Typical dye lasers contain the lasing medium in a transparent vial, and this design allows the dye to be changed to produce different wavelength outputs. A given dye typically has an almost continuous band of laser wavelengths over a range of a few tens of nm. Although such light is not 'monochromatic' in the sense that light from a solid-state or gas laser would be, it is intense and coherent and can be tuned to a specific wavelength of interest. The spectra of some commercially available dyes are shown in figure 2.11, and a simple arrangement for tuning a dye laser to a specific wavelength is shown in figure 2.12.

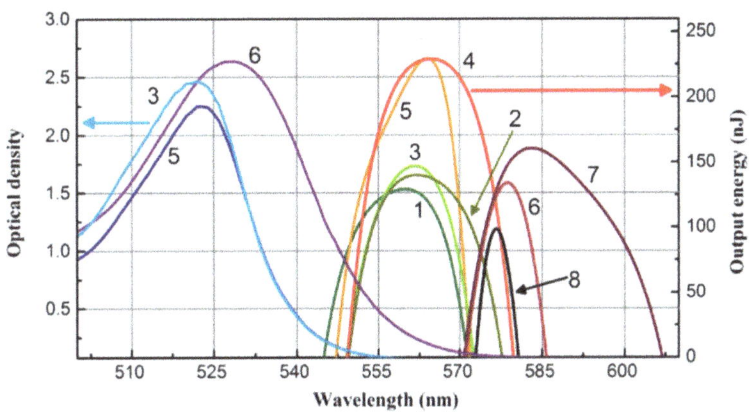

Figure 2.11. The spectral outputs of a dye laser produced using different dyes (indicated by the different numbers in the figure). Reprinted from Burdukova *et al* (2020). © IOP Publishing. Reproduced with permission. All rights reserved.

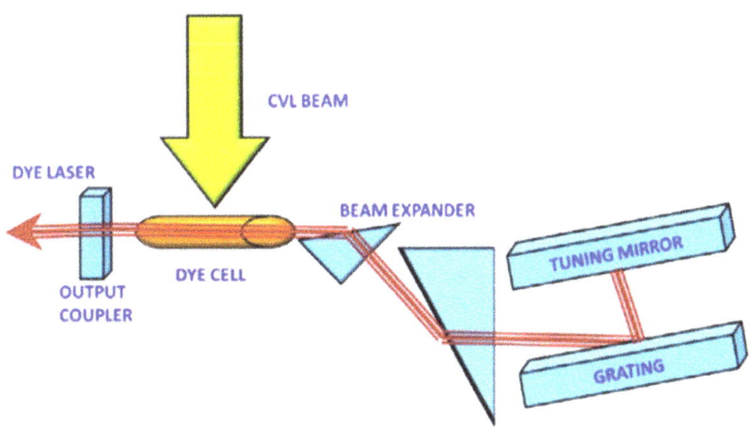

Figure 2.12. A simple method of tuning a dye laser to a specific wavelength, showing the copper vapor laser (CVL) pump beam and the dye laser output beam. The tuning mirror is rotated so that only the desired wavelength is diffracted back through the resonant cavity by the grating. Reprinted from Mishra *et al* (2019). © IOP Publishing. Reproduced with permission. All rights reserved.

2.6 Excimer lasers

For some applications that require ultraviolet laser radiation, another approach can be considered in addition to harmonic generation, namely, lasers which radiate in the ultraviolet. An important example of this type of laser is the excimer laser. The term ‘excimer’ comes from the combination of ‘excited’ and ‘dimer,’ where dimer refers to a molecule of two identical atoms. This name came from a description of the lasing media used in the earliest lasers of this design. Current lasers of this design more commonly use noble gas halides (consisting of molecules made of two different atoms) as the lasing medium. This type of laser is sometimes referred to as an ‘exciplex laser,’ from a combination of ‘excited’ and ‘complex’ laser, although the term excimer laser is most commonly used to refer to both categories of laser.

Excimer lasers are a type of pulsed ultraviolet laser that have the general characteristics of a short wavelength, high energy per pulse, and high average output power. The first excimer lasers were developed in the early 1970s using dimers consisting of two bound noble gas atoms (such as Xe_2). These early lasers were observed to produce intense radiation in the ultraviolet region and this observation prompted considerable research in the field. By the mid-1970s, excimer lasers using a complex consisting of a noble gas atom bound to a halogen atom (such as KrF) has been developed, and this configuration is typically used in contemporary excimer lasers. The wavelengths of some common noble gas–halogen excimer lasers are given in table 2.1. As seen in the table, this radiation is in the ultraviolet and includes transitions in the UV-C (100–280 nm), UV-B (280–315 nm), and UV-A (315–380 nm) portions of the spectrum.

Noble gases do not normally form compounds, either with themselves or with other elements, such as halogens. However, in the excited state, such bonding is possible. This is because the excimer (e.g., Xe_2) or exciplex (e.g., KrF) molecule has an associative (bound) excited state as a result of an attractive molecular interaction but a dissociative (unbound) ground state as a result of a repulsive molecular interaction. The excited state can be formed (for example) by electron pumping or an electrical discharge, and the lasing transition occurs between the excited state and the ground state when a population inversion has been created. When the atoms are in the ground state, they are unbound. When they are pumped up into the excited state, they form a metastable bound state. The spontaneous decay of the excited

Table 2.1. Common molecules used for noble gas–halogen excimer lasers and their lasing wavelengths.

Molecule	Wavelength (nm)
ArF	193
KrCl	222
KrF	248
XeBr	282
XeCl	308
XeF	351

state back to the ground state triggers stimulated laser radiation. When the exciplex molecule returns to the ground state, it quickly dissociates back into individual atoms and this creates the population inversion of the exciplex molecules in the excited state.

Figure 2.13 shows an energy level diagram of a KrF excimer laser in which energy is plotted as a function of interatomic spacing between the Kr and F atoms. Pumping radiation (for example an electron beam) interacts with a Kr atom, placing it in an excited state, Kr^*. This excited noble gas atom interacts with a halogen molecule, F_2, transferring an electron and thereby creating a noble gas positive ion and a negatively charged halogen molecule. The negatively charged halogen molecule dissociates, forming a F^- ion and a neutral F atom. The noble gas ion (Kr^+) and the halide ion (F^-) then form a metastable ionically bound state. Overall, the process is

$$Kr + F_2 \rightarrow Kr_* + F_2 \rightarrow Kr^+ + F_2^- \rightarrow Kr^+F^- + F.$$

The interatomic distance in the metastable molecule is determined by the location of the excited-state energy minimum. When a molecule in the excited state undergoes a transition to the ground state, thereby emitting a photon, it produces the stimulated decay process that results in the emission of a laser pulse at the

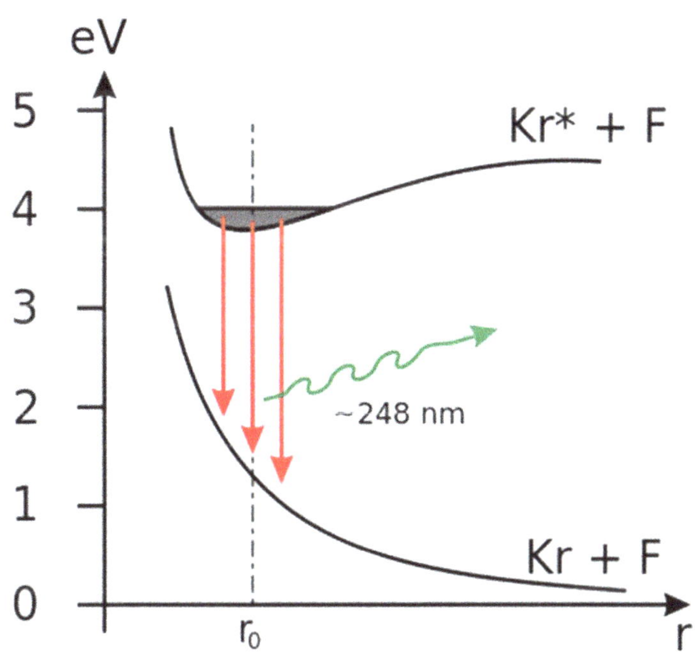

Figure 2.13. Excited (exciplex) and ground-state energies of Kr and F atoms as a function of the distance between the atoms. Public domain by Cepheiden. This [Schematic representation of the electron transition in a KrF laser from the excited to the stable state] image has been obtained by the author from the Wikimedia website, where it is stated to have been released into the public domain. It is included within this book on that basis. (Vertical axis relabeled and 'Elektron' label removed.)

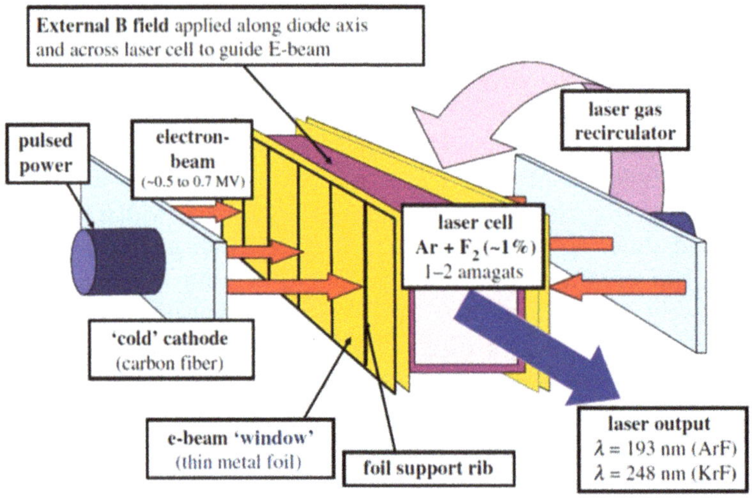

Figure 2.14. An excimer laser design that uses electron-beam pumping. Note: an amagat is a unit of number density defined as the number of ideal gas molecules per unit volume at 1 atm and 0 °C and is equal to 44.615 mol m^{-3}. CC BY 4.0. Reprinted from Obenschain *et al* (2020) © 2020 The Authors.

characteristic wavelength. Since the interatomic interaction in the ground state is repulsive, as there is no local minimum in the energy curve, the molecule quickly dissociates on a timescale of 10^{-12} s.

A typical noble gas–halogen excimer laser using high-energy electron-beam pumping is shown in figure 2.14. The pulsed high-energy electron beam is guided through the lasing channel in order to facilitate interactions with Kr atoms. An important aspect of the operation of the excimer laser is the inclusion of a laser gas recirculator. This allows the heat generated in the lasing medium to be dissipated and permits the laser to be operated at high power for an extended period of time with minimal heating.

References and suggestions for further reading

Basting D and Marowsky G (ed) 2005 *Excimer Laser Technology* (Berlin: Springer)

BP-Aegirsson 2017 Schematic view of the SHG conversion of an exciting wave in a non-linear medium with a non-zero second-order non-linear susceptibility https://commons.wikimedia.org/wiki/File:Schematic_of_the_SHG_conversion_of_an_excited_wave_in_a_non-linear_medium.png

Burdukova O A, Dolotov S M, Petukhov V A and Semenov M A 2020 Tunable polymer dye laser pumped by two 513 nm diodes *Laser Phys. Lett.* **17** 025801

Cepheiden 2008 Schematic representation of the electron transition in a KrF laser from the excited to the stable state https://commons.wikimedia.org/wiki/File:Übergang_KrF-Excimer.svg

Dunlap R A 2021 *Energy from Nuclear Fusion* (Bristol: IOP Publishing) https://iopscience.iop.org/book/mono/978-0-7503-3307-8

Elliott D J 2014 *Ultraviolet Laser Technology and Applications* (New York: Academic)

Ewing J J 2000 Excimer laser technology development *IEEE J. Sel. Top. Quantum Electron.* **6** 1061–71

Mishra S K, Rana P, Sahoo S P, Kawade N O and Rawat V S 2019 Characterization of dye cells for a high-repetition-rate pulsed dye laser by particle image velocimetry (PIV) *Laser Phys.* **29** 065001

Obenschain S P, Schmitt A J, Bates J W, Wolford M F, Myers M C, McGeoch M W, Karasik M and Weaver J L 2020 Direct drive with the argon fluoride laser as a path to high fusion gain with sub-megajoule laser energy *Phil. Trans. R. Soc.* A **378** 20200031

Presciutti A, Asdrubali F, Marrocchi A, Broggi A, Pizzoli G and Damiani A 2014 Sun simulators: development of an innovative low cost film filter *Sustainability* **6** 6830–46

Shankarling G S and Jarag K J 2010 Laser dyes *Resonance* **15** 804–18

Silfvast W T 1996 *Laser Fundamentals* (Cambridge: Cambridge University Press)

Strome F C and Webb J P 1971 Flashtube-pumped dye laser with multiple-prism tuning *Appl. Opt.* **10** 1348–53

Svelto O 1998 *Principles of Lasers* 4th edn (New York: Springer)

Wilson J and Hawkes J F B 1987 *Lasers: Principles and Applications* (New York: Prentice-Hall)

XuPanda 2017 Energy levels in a He-Ne Laser https://commons.wikimedia.org/wiki/File:HeNe_Laser_Levels.png

Chapter 3

Semiconducting lasers

3.1 Introduction

Traditional lasers based on solid, liquid, or gaseous media that contain atoms, ions, or molecules that exhibit lasing transitions have been important for a number of very significant scientific and industrial applications, as discussed in chapter 4; however, it was the development of semiconducting lasers, as described in this chapter, that led to the widespread use of lasers in applications related to consumer electronics and communication. The basic physics of semiconducting materials and the method used to produce coherent stimulated radiation from a semiconducting junction are summarized in this chapter.

3.2 Semiconductor physics

Semiconducting lasers produce coherent laser radiation as a result of transitions between the conduction band and the valence band of semiconducting materials. The formation of a bandgap in a solid can be explained by the interaction between the electrons in the solid and the Coulomb potential of the ions. We begin by looking at the behavior of a free electron. The relationship between the electron energy, E, and momentum, p, is given by

$$E = \frac{p^2}{2m_e}, \tag{3.1}$$

where m_e is the electron mass. The de Broglie wavelength of the electron, λ, is given in terms of its momentum and the reduced Planck constant, $\hbar$, as

$$\lambda = \frac{2\pi\hbar}{p}. \tag{3.2}$$

doi:10.1088/978-0-7503-5482-0ch3

By combining equations (3.1) and (3.2), we obtain

$$E = \frac{\hbar^2 k^2}{2m_e}, \tag{3.3}$$

where the wavevector, k is defined as

$$k = \frac{2\pi}{\lambda} = \frac{p}{\hbar}.$$

The expression in equation (3.3) is the dispersion relation for the free electron and shows that the relationship between the electron energy and its wavenumber is quadratic.

In a crystalline solid, the valence electrons are subject to the periodic Coulomb potential of the ions on the lattice sites. If the lattice spacing is a, then when a half-integer number of electron wavelengths is equal to a, the electron density is at a minimum at the ion locations. This occurs for

$$k = \frac{\pi n}{a}, \tag{3.4}$$

where n is an integer. When this condition is met, the electrons form standing waves and the slope of the dispersion relation, which is the group velocity of the electron waves, goes to zero. This means that energy gaps (or bandgaps) form in the electron energy relation for wavevectors that satisfy equation (3.4), and this is illustrated in figure 3.1. The energy gap, E_g in the figure, represents a range of energies for which no allowed electron states occur. In this simple mode, the electron energy levels of the system are divided into bands which are separated by gaps, for which it can be shown that the number of allowed energy levels in each band is equal to the number of atoms in the material. Since electrons can exist in two different spin states, each band can accommodate two electrons per atom.

This very simple picture of electron behavior in solids is referred to as the nearly free electron model. Further details of the basic theory of band formation in solids are provided in Dunlap (1988).

To begin our overview of semiconducting lasers, we first consider the effects of impurities on the electrical properties of semiconductors. We then consider the properties of semiconducting junctions.

As a simple example, we consider a generic valence four solid. Here, the ions are covalently bonded with tetrahedral coordination. The bonds in the three-dimensional tetrahedral structure can be illustrated by a square lattice in two dimensions, as shown in figure 3.2. In this analog, neighboring atoms each share two valence electrons.

Semiconducting materials that are used in practical devices almost always contain a controlled concentration of impurities, referred to as doping. We can consider a simple example of two different situations: one in which the impurities have a valence of five, referred to as donors, and one in which the impurities have a valence of three, referred to as acceptors. These two situations are illustrated in figures 3.2(a)

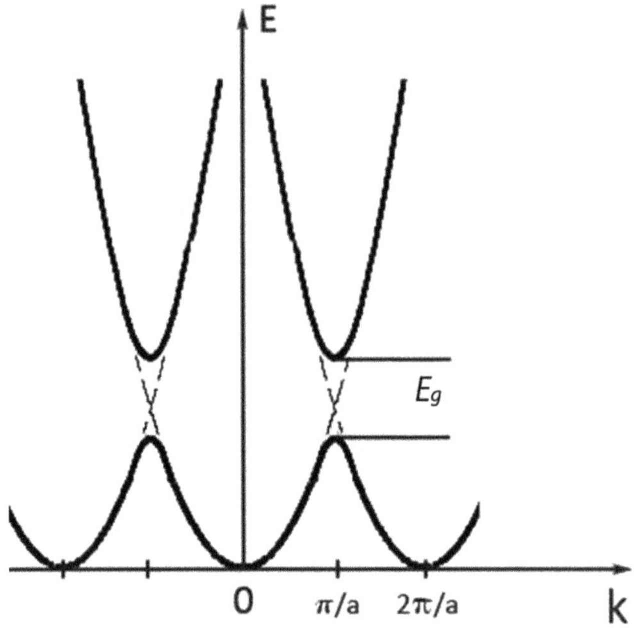

Figure 3.1. The formation of energy gaps in electron dispersion relation as a result of interaction with a periodic Coulomb potential. This representation of the band structure is referred to as the repeated zone scheme. This [Electronic band structure in the nearly free electron picture] image has been obtained by the author from the Wikimedia website where it was made available under a CC BY-SA 3.0 licence. It is included within this book on that basis. It is attributed to Killkoll (label of energy gap changed to E_g).

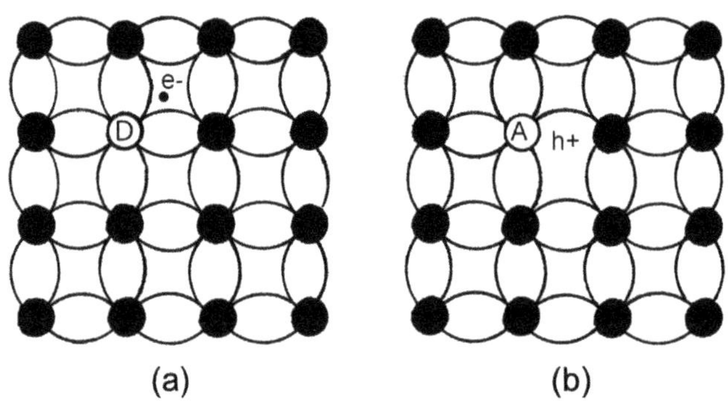

Figure 3.2. A two-dimensional representation of tetrahedral bonding in a valence four semiconductor with (a) a valence five impurity, i.e. a donor, D, and (b) a valence three impurity, i.e. an acceptor, A. Adapted from Dunlap R A (2018) with permission of Morgan & Claypool Publishers.

and (b), respectively. In the case of a valence five impurity, each impurity atom has one more electron than the host atoms. Four of its valence electrons contribute to bonding, and the fifth electron is free to move around in the material. This situation is illustrated in figure 3.2(a). A valence four semiconducting material that is doped with valence five impurities is referred to as an n-type semiconductor. In the case of a valence three impurity, there is one electron missing and this is represented by the missing bond in figure 3.2(b). This missing bond behaves like a positive charge. It is referred to as a hole and can propagate throughout the material. A valence four semiconducting material that is doped with valence three impurities is referred to as a p-type semiconductor. The most common valence four semiconducting host material, silicon, can be doped with valence five impurities such as phosphorus, arsenic, or antimony to create an n-type semiconducting material. Silicon can be doped with valence three impurities, such as boron, gallium, indium, or aluminum, to create a p-type semiconducting material.

We can view the impurity state in a semiconductor using the dispersion relation. This illustrates the band structure, and a very simplified picture is shown in figure 3.3. The lowest energy band in the figure represents the energy levels of the two lowest-energy valence electrons and is referred to as the valence band. The highest energy band in the figure represents the energy levels of the next two highest-energy electrons and is referred to as the conduction band. Filled lower energy bands are not shown in the figure, as they do not directly contribute to the electrical properties of the material.

In the case of a valence five impurity, four of the valence electrons are sufficient to fill up the valence band, and the fifth valence electron goes into the conduction band. Therefore, in figure 3.3(a), there is one electron in the conduction band for each valence five impurity atom in the sample. In the case of a valence three impurity, the number of valence electrons is not sufficient to fill up the valence band, and this leaves a hole at the top of the valence band, as shown in figure 3.3(b). Therefore, there is one hole in the valence band for each valence three impurity atom in the sample.

Thus, the introduction of impurities with the appropriate electronic configuration provides a means of creating electron states in the conduction band or hole states in

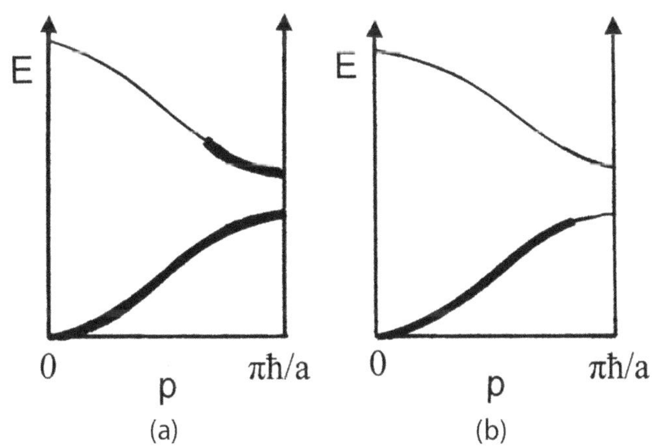

Figure 3.3. Dispersion relation showing the band structure and occupied states for a semiconductor with (a) donor impurities and (b) acceptor impurities. This representation of the band structure is referred to as the reduced zone scheme. Note $p = \hbar k$.

the valence band of a semiconducting material, thereby creating an n-type material or a p-type material, respectively. As these electrons and holes are free to move throughout the material, they can contribute to the electrical conductivity of the material by responding to any applied electric fields, as described in further detail below.

Since semiconducting materials are most commonly used at around room temperature, it is important to consider thermal effects. Thermal energy can cause electrons from the valence band to be excited up into the conduction band, leaving hole states behind them in the valence band. Thus, in a piece of n-type doped semiconductor there are a lot of free electron carriers produced by donor impurities. There are also a smaller number of electron carriers produced by thermal excitations and an equal number of hole carriers that are formed. In an n-type semiconductor, the electrons are called the majority carriers and the holes are called the minority carriers. In a p-type semiconductor the roles of electrons and holes are interchanged.

The behaviors of electron and hole charge carriers in a semiconductor can be viewed somewhat more quantitatively by looking at some basic physics. We begin with a free electron at rest in a piece of semiconducting material, as illustrated in figure 3.4. If we close the switch in the diagram and supply a potential difference, V, between the two ends of the material, then an electric field, E, will be present, where

$$E = \frac{V}{l}.$$

This will exert a Lorentz force, F, on the negatively charged electrons,

$$F = -eE = -\frac{eV}{l}.$$

Newton's law gives the acceleration of the electron as

$$a = \frac{F}{m_e} = -\frac{eV}{m_e l},$$

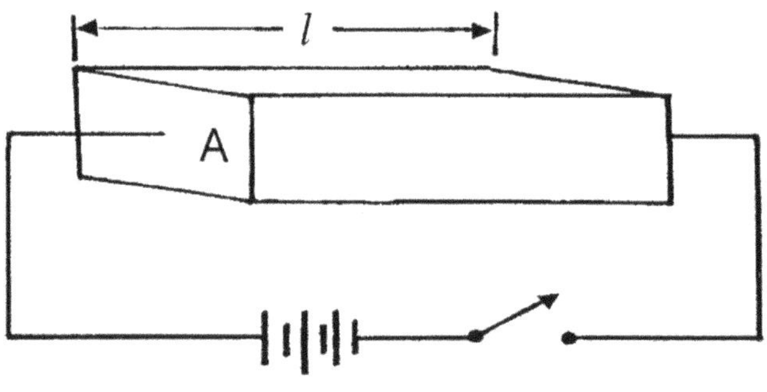

Figure 3.4. An experiment for measuring the conductivity of a sample of material with cross-sectional area A and length l.

where m_e is the electron mass. At a time t after the switch is closed, the electron velocity will be

$$v = at = -\frac{eVt}{m_e l}.$$

The minus sign merely means that the electrons will flow in the direction opposite to that of the electric field because of their negative charge. The electric current flowing through the sample is given by the charge transported per unit time. If n_e is the free electron density in the material, then the current can be expressed as

$$I = -en_e vA = \frac{e^2 n_e VAt}{m_e l}, \tag{3.5}$$

where A is the cross-sectional area of the sample, as shown in the figure. Using Ohm's law, $V = IR$, we can rewrite equation (3.5) as

$$V = I\left(\frac{m_e l}{e^2 n_e At}\right)$$

or

$$R = \frac{m_e l}{e^2 n_e At},$$

where R is the resistance. Since the resistivity is $\rho = AR/l$ and the conductivity is $\sigma = 1/\rho$, the conductivity may be written as

$$\sigma = \frac{e^2 n_e}{m_e}t.$$

Some of this makes sense, since the conductivity is proportional to the charge on the carriers and their density, and it is inversely proportional to their mass. However, it does not make sense that the conductivity increases linearly with time after the switch is closed. Free electrons in the material are not really free, as noted in the discussion above, but interact with other charges and phonons (i.e. quantized lattice vibrations). Therefore, the electrons travel for some time (the mean free time, τ_e) before interacting substantially, at which time they have to start accelerating all over again. Therefore, the conductivity reaches some equilibrium value which is written as

$$\sigma_e = \frac{e^2 n_e}{m_e}\tau_e = en_e\mu_e,$$

where $\mu_e = e\tau_e/m_e$ is defined as the mobility of the electrons.

We can repeat this derivation for the hole carriers, given that their number density is n_h and their mobility may be defined in terms of their own mean free time, τ_h, and effective mass, m_h, so that $\mu_h = e\tau_h/m_h$ and the conductivity of the holes is thus

$$\sigma_h = en_h\mu_h.$$

We note that the sign of the charge on the holes is positive and the current flows in the opposite direction to that of the electrons. However, the Lorentz force also acts in the opposite direction. This means that the electrons and holes travel in opposite directions but contribute to a current in the same direction. This means that we can write the total conductivity as

$$\sigma = \sigma_e + \sigma_h = en_e\mu_e + en_h\mu_h.$$

While the above discussion describes the behaviors of n-type and p-type semiconducting materials, most semiconducting devices, including semiconductor lasers, utilize junctions between these two types of material. The next section begins with a discussion of what happens when n-type and p-type semiconductors are placed in electrical contact.

3.3 Semiconducting junctions

In order to understand what happens when semiconducting materials of different types form junctions, it is first important to know where all of the charges in an n-type and in a p-type semiconductor are located and how they behave. In a piece of n-type material, we have neutral host atoms that are combined with neutral donor atoms. Thus, the doped semiconductor must be neutral overall. There are negatively charged electrons that are freed from donor atoms. There are an equal number of donor atoms which have lost an electron and have become positive ions. There are negatively charged electron that are thermally excited and there are an equal number of positively charged holes that are created in this process. Thus, the electrical neutrality is maintained, but the behaviors of the positive and negative charges are different. There are the majority negatively charged mobile electron carriers that are the result of both donor electrons from the impurities and thermally excited electrons. There are the minority positively charged mobile holes that are the result of thermal excitations. Finally, there are the positively charged impurity ions which are part of the lattice and are not free to move. In a piece of p-type material, the charges are just the opposite.

Another feature of a semiconducting material that is important to realize is that semiconducting materials are ohmic, that is, they obey Ohm's law. It is only when junctions are formed from different types of semiconducting material that non-ohmic behavior results. The simplest junction is a single piece of n-type material joined to a single piece of p-type material as shown in figure 3.5. This is the basic design of a diode. The identities of the charges in the figure are described in the caption. We begin with a discussion of the behavior of the majority carriers, that is, the electrons in the n-type material and the holes in the p-type material.

There are two aspects of the majority carriers in the system that need to be considered: (1) the concentration of the majority carriers on each side of the junction and (2) the spatial distribution of the carriers. With respect to the first point, the majority electrons in the n-material see a much smaller electron concentration in the p-material on the other side of the junction. This concentration gradient across the junction provides a driving force for electron diffusion from the high

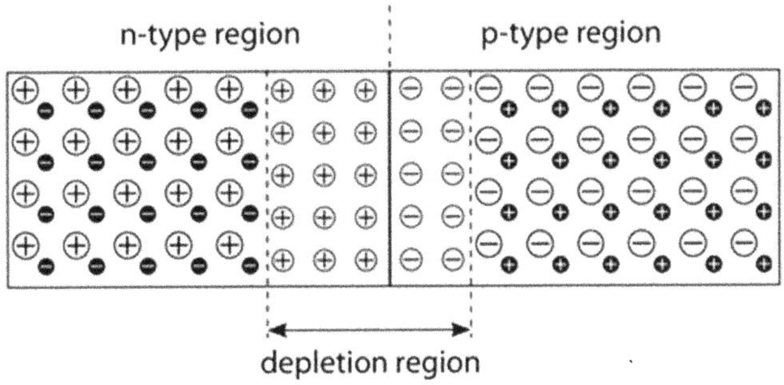

Figure 3.5. A p–n semiconducting junction or diode. Donor ions and acceptors are denoted by large open circles (+) and (−), respectively. Electron and hole carriers are shown by small black circles (−) and (+), respectively. CC BY-SA 2.5. This [p–n Unpolarized junction] image has been obtained by the author from the Wikimedia website where it was made available under a CC BY-SA 2.5 licence. It is included within this book on that basis. It is attributed to Wojciech mula (labels translated to English).

concentration in the n-material to the low concentration in the p-material in an attempt to distribute the electrons evenly. Similarly, the majority hole carriers in the p-material see a lower hole concentration in the n-material on the other side of the junction, and this provides a driving force for hole diffusion from the p-material to the n-material. As soon as this diffusion begins, the n-material loses negatively charged electrons and gains positively charged holes, while the p-material loses positively charged holes and gains negatively charged electrons. This creates an excess of positive charge on the n-side of the junction and an excess of negative charge on the p-side of the junction, thus giving rise to the formation of an electric field pointing from left to right across the junction in the figure. This field opposes the diffusion of the electrons and holes. An equilibrium situation is be set up in which the resulting electric field is just sufficient to compensate for the driving force for diffusion that results from concentration gradients.

The spatial distribution of majority carriers can be viewed in the following way. The negatively charged electrons in the n-region see the negative charge of the fixed acceptor impurities on the other side of the junction and are repelled. As a result, they are redistributed in the n-region away from the junction, as shown in the figure. Similarly, the holes in the p-region see the positive charge of the fixed donor impurities in the n-region and are also repelled away from the junction as shown. If we look at the majority carrier distribution as shown in figure 3.5, we readily see that there are lots of charge carriers in the region away from the junction and virtually no charge carriers in the region just on either side of the junction. This region that is depleted of charge carriers is called the depletion region or depletion layer. If we were able to measure the electrical conductivity in the various regions of the system, we would find that the regions far from the junction would have a fairly high conductivity because of the high density of free carriers. The depletion region, on the other hand, would have a very low conductivity because of the lack of mobile charge carriers in the region. As a result, the electric field, and hence the change in the electric potential, occurs primarily across the depletion region.

Finally, we need to look briefly at the behavior of the minority carriers. These are the electron and hole carriers that are created by thermal excitations. If a minority hole carrier in the n-region drifts (because of thermal motion) into the depletion region, it experiences a force due to the presence of the electric field in this region. The electric field pushes the hole across the junction to the p-region, where it becomes a majority carrier. Similarly, a minority electron on the p-side of the junction which drifts into the depletion region is carried by the electric field across the junction to the n-region, where it becomes a majority carrier. These minority currents across the junction unbalance the majority charge carrier configuration. As a result, majority charge carriers flow across the junction to compensate for this imbalance. Since no net current flows in or out of the diode, all of these majority and minority currents must cancel.

We can view the equality of majority and minority currents quantitatively as follows. The details of the two majority and two minority currents are shown in table 3.1. The total current, I_{total}, must be equal to zero:

$$I_{\text{total}} = I_1 + I_2 + I_3 + I_4 = 0. \tag{3.6}$$

We also know that the electron currents must cancel each other, and the hole currents must cancel each other, so that

$$I_1 = -I_2 \tag{3.7}$$

and

$$I_3 = -I_4. \tag{3.8}$$

A convenient way of picturing the p–n junction is shown in figure 3.6, in which energy is plotted on the vertical axis and spatial position is plotted on the horizontal axis. The figure shows the locations of the valence and conduction bands and indicates the filled and vacant states in each. A very important aspect of the energy levels that is seen in this diagram is that the levels in the p-region are shifted upward (to a higher energy) from those in the n-region. This shift is caused by the electric field that is set up across the junction by the redistribution of mobile charge carriers.

If a diode is connected to an external voltage source, then the interesting features of a semiconducting junction become apparent. As the two sides of the diode are different in terms of the polarity of the charge carriers, there are two different ways

Table 3.1. Definitions of the majority and minority currents across the diode junction. A positive current is defined as a positive flow of charge from left to right (i.e. n-type to p-type).

Current	Type	Carrier	Direction	Sign
I_1	Minority	Electrons	$\mathrm{p} \rightarrow \mathrm{n}$	(+)
I_2	Majority	Electrons	$\mathrm{n} \rightarrow \mathrm{p}$	(−)
I_3	Majority	Holes	$\mathrm{p} \rightarrow \mathrm{n}$	(−)
I_4	Minority	Holes	$\mathrm{n} \rightarrow \mathrm{p}$	(+)

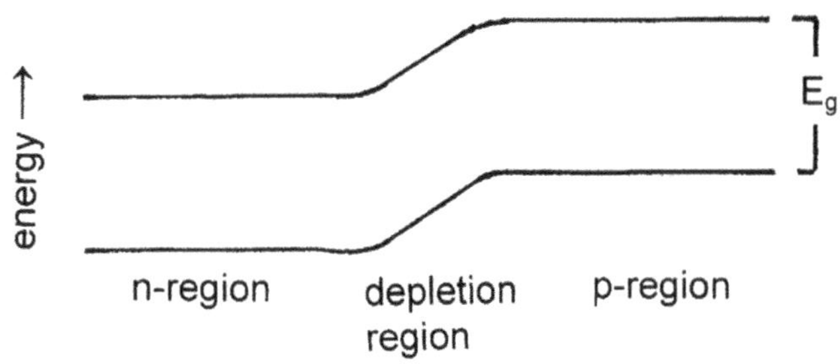

Figure 3.6. Energy bands in a diode.

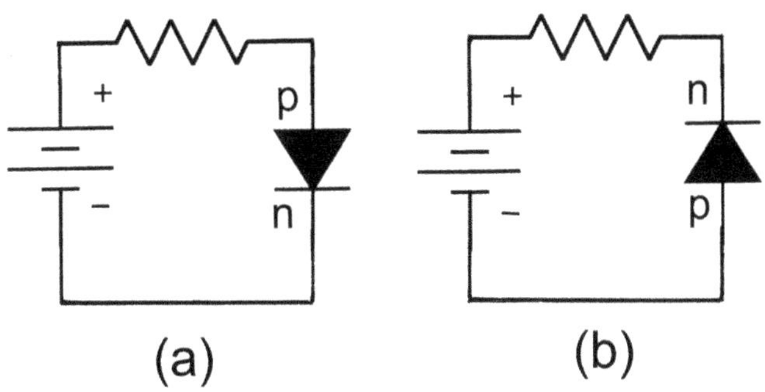

Figure 3.7. (a) A forward biased diode and (b) a reverse biased diode. The symbol for the diode is a triangle with a bar, in which the triangle points from the p-side to the n-side of the junction.

in which the external voltage can be applied to the device. These two ways are shown in figure 3.7. The symbol for the diode is a triangle with a bar. The triangle points from the p-side to the n-side of the diode, as shown in the figure. The conditions shown in the figure are referred to as the forward bias case and the reverse bias case, corresponding to the application of the positive voltage to the p-side of the diode or the n-side of the diode, respectively.

The application of an external voltage to the diode produces an electric field across the device. This electric field either adds to the internal field or subtracts from it. Recall that the internal field in the device is a result of excess positive charge on the n-side and excess negative charge on the p-side. In the forward bias case, we see that the applied field opposes the internal field in the diode; conversely, in the reverse bias condition, the external field adds to the internal field. If we refer to figure 3.6, we see that the change in the energy levels across the junction that results from the charge redistribution that creates the internal field inhibits majority charge carrier movement across the junction. In the forward bias case, the external voltage reduces this change in energy levels and makes it easier for majority carriers to flow across the junction. In the reverse bias case, just the opposite happens, and it is more difficult for the majority carriers to flow. The minority currents are not affected by the external field, as these result from thermal motion. So, from table 3.1, we see that the external field affects currents I_2 and I_3. It can be shown that in the presence of an

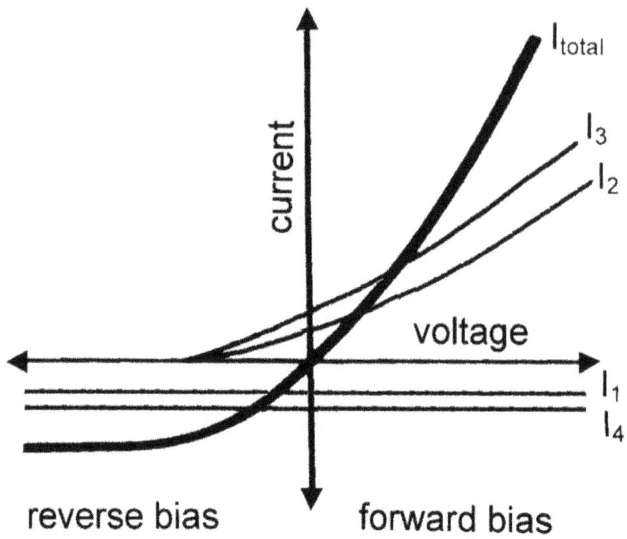

Figure 3.8. The total current and individual currents flowing through a diode as a function of the bias voltage.

applied voltage, V, the majority currents for the forward (+) and reverse (−) bias conditions become

$$I_2^{\pm} = I_2 \exp\left(\pm\frac{eV}{k_B T}\right) \tag{3.9}$$

and

$$I_3^{\pm} = I_3 \exp\left(\pm\frac{eV}{k_B T}\right), \tag{3.10}$$

respectively. Combining equations (3.6), (3.7), and (3.8) with equations (3.9) and (3.10) gives

$$I_{\text{total}} = (I_1 + I_4) \cdot \left[1 - \exp\left(\pm\frac{eV}{k_B T}\right)\right]. \tag{3.11}$$

Equation (3.11) is called the rectifier equation or diode equation and gives the total current flowing through the diode as a function of the applied voltage. Figure 3.8 shows the voltage dependence of the individual currents and the total current through the diode.

3.4 Light-emitting diodes and semiconductor lasers

The simplest method of producing light from a semiconducting junction is by forward biasing (as illustrated in figure 3.7(a)). In this case, electrons are injected from the power supply into the n-side of the diode and holes are injected into the p-side of the diode. The forward bias increases the majority carriers across the junction and increases the forward current as shown in figure 3.8. When the

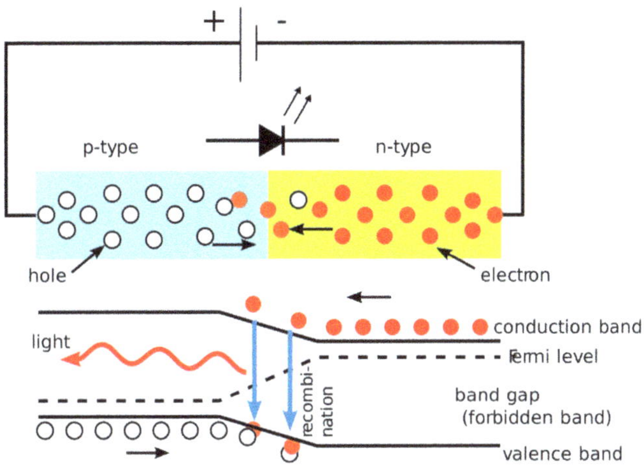

Figure 3.9. The production of photons by the recombination of electrons and holes in the depletion region of a forward biased p–n junction. CC BY-SA 2.5. This [Schematic diagram of Light–Emitting Diode (LED)] image has been obtained by the author from the Wikimedia website where it was made available under a CC BY-SA 2.5 licence. It is included within this book on that basis. It is attributed to S-kei.

electrons from the n-side and the holes from the p-side enter the depletion region, they can recombine; in this process, an electron from the conduction band effectively falls across the energy gap to fill a hole in the valence band. This process is illustrated in figure 3.9. The change in electron energy can be liberated in the form of a photon, thereby emitting light from the depletion region.

The energy levels in the semiconductor valence and conduction bands are quantized but are very close together. The electrons in the conduction band can be in a variety of different levels, and the holes in the valence band with which they recombine can also be in a variety of levels. However, it is most likely that electrons near the conduction band edge will recombine with holes near the valence band edge, meaning that the energy that is released to produce the photon is close to the energy of the bandgap.

Figure 3.10 shows the spectrum of a red light-emitting diode (LED). While this spectrum is not broadband, the range of wavelengths emitted is much larger than for (e.g.) a gas discharge tube. This is because of the nature of the transitions between the conduction and valence bands as described above.

The wavelength of the light emitted by an LED can be varied by choosing a semiconducting material with an energy gap that is commensurate with the desired wavelength. Table 3.2 lists common semiconducting materials that are used for LEDs of different colors. Most of these materials, e.g. GaP, are referred to as III–V semiconductors. The semiconductors described in section 3.2 were valence four semiconductors (such as Si). Equal proportions of a valence three material (e.g. Ga) and a valence five material (e.g. P) act as a valence four semiconductor. The equivalent of a doped material can be created by adjusting the composition; thus, materials with more of the valence three component act as p-type materials and materials with more of the valence five component act as n-type materials.

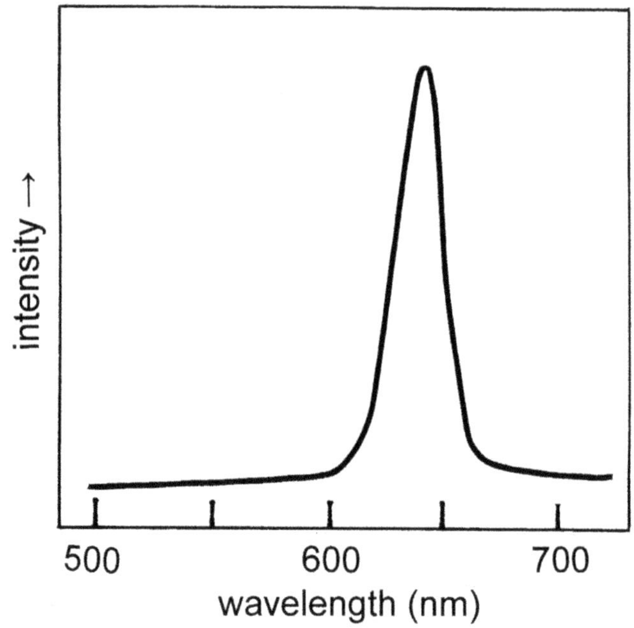

Figure 3.10. The spectrum of a red LED.

Table 3.2. Some common LED materials used to produce light of different colors.

Color	Wavelengths (nm)	Semiconducting materials
Infrared	$\lambda > 760$	GaAs
		GaAlAs
Red	$610 < \lambda < 760$	GaAlAs
		GaP
		GaAlInP
Orange	$590 < \lambda < 610$	GaAsP
		GaP
		GaAlInP
Yellow	$570 < \lambda < 590$	GaAsP
		GaP
		GaAlInP
Green	$500 < \lambda < 570$	GaP
		GaAlP
		GaAlInP
Blue	$450 < \lambda < 500$	GaInN
		ZnSe
Violet	$400 < \lambda < 450$	GaInN
Ultraviolet	$\lambda < 400$	AlN
		BN
		C (diamond)
		GaAlN
		GaAlInN
		GaInN

These materials allow for the construction of p–n junctions. Adjusting the composition also results in variations in the energy gap and hence the wavelength of the LED light that is produced.

For many applications (e.g. household lighting), white light is preferred rather than light of a single color. White LEDs can be created using two methods. A straightforward method is to combine LEDs of different colors (i.e. red, green, and blue) so that the eye perceives the light as white. However, a simpler technique is to use light from a short-wavelength (blue) LED to irradiate a phosphoric material, which then re-irradiates photons in a broad band over the visible region, as shown in figure 3.11.

While LEDs radiate light that is more or less a single color, they are not as monochromatic as lasers, and the light they produce is not coherent. In order to produce the equivalent of laser radiation from an LED, it is necessary to construct a resonant cavity, so that the light is amplified, and to create something equivalent to a population inversion, so that the radiation is stimulated rather than spontaneous.

A simple design for a resonant cavity in a semiconducting device is illustrated in figure 3.12. Partially and fully reflecting surfaces are made on the ends of a semiconductor so that light produced in the active junction region can be amplified. In order to improve the efficiency of the laser diode, more complex junction geometries are generally utilized which involve several layers of semiconducting materials with differing impurity levels. Stimulated emission results when a sufficient number of majority carriers are injected into the depletion region. Figure 3.13 shows

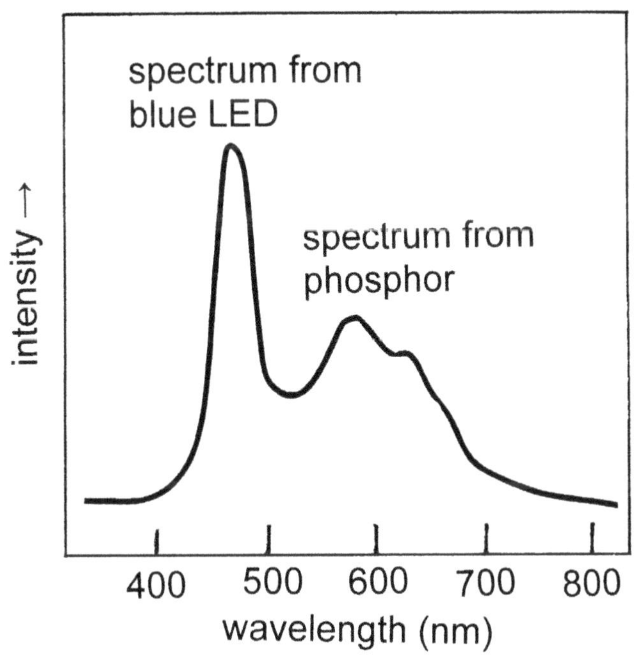

Figure 3.11. The optical spectrum of a white LED created by a blue LED incident on a broadband phosphor.

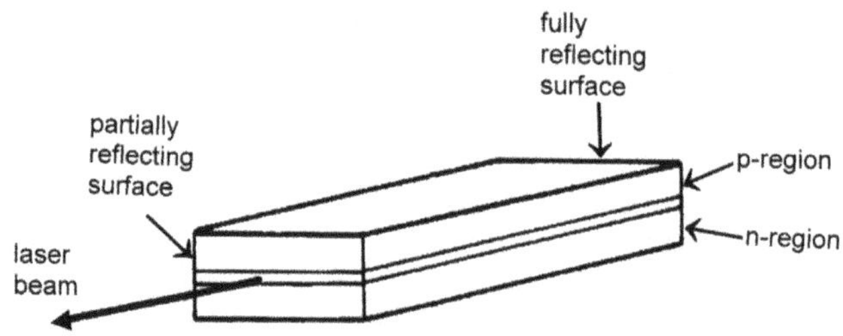

Figure 3.12. The design of a laser diode.

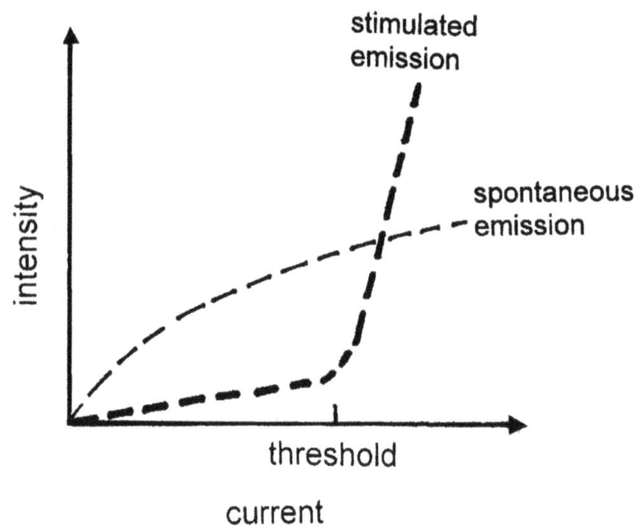

Figure 3.13. The energy outputs of a LED and a laser diode as a function of the forward bias current.

the light output as a function of the forward bias current for a typical LED and a typical laser diode. It can be seen that above a threshold current, the laser diode begins to produce intense stimulated emission and becomes much brighter than the LED.

The spectrum of a laser diode is illustrated in figure 3.14. These results show the appearance of coherent radiation, as illustrated in figure 3.13, and the formation of cavity modes. The spectrum in figure 3.14 can be compared with the spectrum of a He–Ne laser in figure 1.14. Although the general features of these two spectra are similar, there is an important and substantial difference; the range of wavelengths is much greater for the laser diode than for the He–Ne laser. There are two important factors that are responsible for this behavior. First, the energy levels of the electrons and holes, as mentioned above, are not single, well-defined levels but have a broader range of transition energies than for the atomic energy levels in a gas. Second, the resonant cavity is typically a fraction of a millimeter in length rather than tens of centimeters. The first feature means that the overall distribution of wavelengths (i.e.

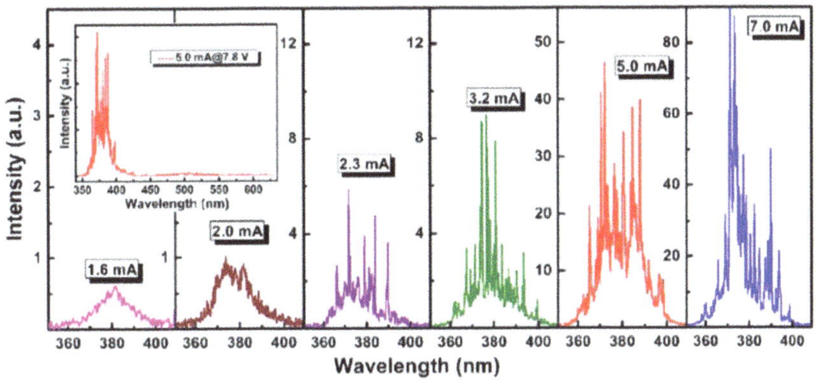

Figure 3.14. The spectra of a laser diode comprised of an epitaxial ZnO/MgO–core/shell nanowall structure as a function of current. CC BY-SA 4.0. Reprinted from Shi *et al* (2014) Copyright © 2014, The Authors.

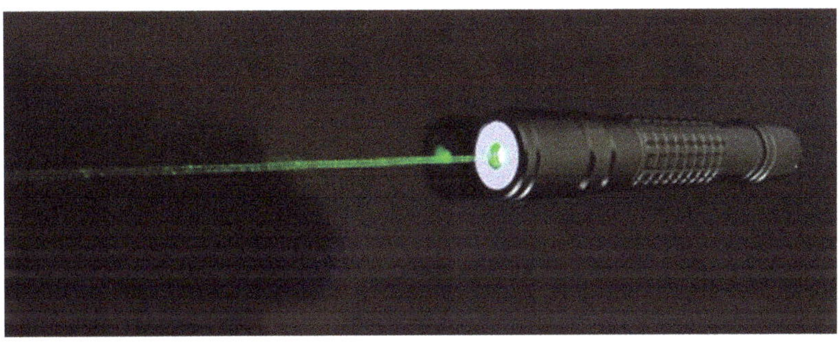

Figure 3.15. A green laser pointer utilizing frequency doubling. This [Q-LINE lasers] image has been obtained by the author from the Wikimedia website where it was made available (Netweb01) under a CC BY 3.0 licence. It is included within this book on that basis. It is attributed to Kreuzlinienlaser für den Hausbau/Heimwerker. (Image cropped.)

the envelope in figure 1.13) is much greater for the semiconducting laser than for the atomic laser. The second feature means that the spacing of the resonant modes is much greater for the semiconducting laser. This can be seen in the $1/L$ dependence of equation (1.8).

An interesting and common application of laser diodes is in laser pointers. Traditional red laser pointers emit at about 650 nm and utilize a simple and inexpensive GaInP/AlGaInP diode. Green laser pointers (see figure 3.15) have become common in recent years. New designs incorporate an InGaN laser diode that emits at about 520 nm, although traditional green laser pointer designs utilize an interesting technology that incorporates concepts from both this chapter and the previous chapter. Figure 3.16 shows a basic diagram of a diode-pumped solid-state (DPSS) laser, sometimes called a diode-pumped solid-state frequency-doubled (DPSSFD) laser. A laser diode (typically AlGaAs) emits in the near-infrared at around 808 nm. This near-infrared laser radiation pumps a neodymium-doped crystal (typically neodymium-doped yttrium orthovanadate ($Nd:YVO_4$)) which

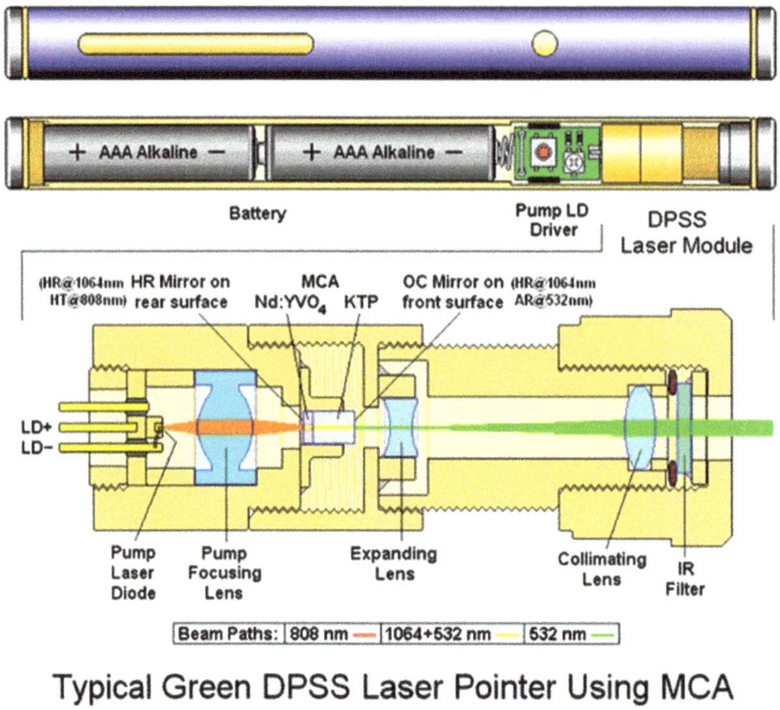

Figure 3.16. A diagram of a typical green laser pointer. Courtesy of Goldwasser (1994–2004).

produces laser light in the deeper infrared at around 1064 nm. This infrared laser radiation is incident on a crystal of potassium titanyl phosphate ($KTiOPO_4$), commonly known as KTP, which is a nonlinear optical material that serves as a second-harmonic generator (see section 2.3). The output of the second-harmonic-generating crystal is at 532 nm combined with a residual 1064 nm component. The infrared component is filtered out by an infrared filter, which leaves green laser radiation at 532 nm.

References and suggestions for further reading

Dunlap R A 1988 *Experimental Physics—Modern Methods* (New York: Oxford University Press)

Dunlap R A 2018 *Novel Microstructures for Solids* (San Rafael, CA: Morgan & Claypool)

Hecht J 1992 *The Laser Guidebook* 2nd edn (New York: McGraw-Hill)

Killkoll 2008 Electronic band structure in the nearly free electron picture https://commons.wikimedia.org/wiki/File:BandstructureNFE.PNG

Meyer T, Braun H, Schwarz U T, Tautz S, Schillgalies M, Lutgen S and Strauss U 2008 Spectral dynamics of 405 nm (Al,In)GaN laser diodes grown on GaN and SiC substrate *Opt. Express* **16** 6833–45

Netweb01 2010 Q-LINE lasers https://commons.wikimedia.org/wiki/File:Laser_pointers.jpg

Goldwasser S M 1994–2004 Components of Typical Green DPSS Laser Pointer (grnptr1.jpg) http://www.walshcomptech.com/repairfaq/sam/glpmca1.gif

Shi Z, Zhang Y, Cui X, Zhuang S, Wu B, Dong X, Zhang B and Du G 2014 High-temperature continuous-wave laser realized in hollow microcavities *Sci. Rep.* **4** 7180

S-Kei 2011 Schematic diagram of Light Emitting Diode (LED) https://commons.wikimedia.org/wiki/File:PnJunction-LED-E.svg
Svelto O 1998 *Principles of Lasers* 4th edn (New York: Springer)
Wilson J and Hawkes J F B 1987 *Lasers: Principles and Applications* (New York: Prentice-Hall)
Wojciech mula 2005 Unpolarized p-n junction https://commons.wikimedia.org/wiki/File:Zlacze_pn-niespolaryzowane.svg

Chapter 4

Laser applications

4.1 Introduction

There are an enormous number of applications for lasers that are based on one or more of the fundamental properties of laser light: intensity, monochromaticity, and coherence. The details of the application requirements determine the most suitable type of laser. This chapter reviews only a few of the numerous commercial, industrial, and scientific applications of lasers that make use of these laser properties. Part II of this book presents a description of the application of lasers to the cooling and trapping of individual atoms.

4.2 Communications

A simple fiber optic cable consists of a glass core surrounded by a cladding made from glass with a different index of refraction. Light propagates along the fiber in the core as a result of total internal reflection caused by the cladding. More complex designs involve graded-index fibers, in which the light is gradually bent back to the center of the fiber rather than by abrupt reflection at an interface.

Fiber optic communication systems use a modulated LED or laser diode as a source of electromagnetic radiation with which to transmit signals. Figure 4.1 shows the basic design of such a system. An electrical signal is used to modulate the output of the light source. The resulting optical signal is coupled to a fiber optic cable for transmission. Relay stations or regenerators can be used to extend the range or bandwidth of the transmission, as discussed below. The receiver consists of a photodiode that converts the modulated light signal back into an electrical signal. The photodiode operates in the reverse manner to that of a light-emitting diode. That is, photons which are incident on the depletion region of the diode excite electrons from the valence band to the conduction band, and the electron–hole pairs thus created constitute a current through the diode, which can then be measured.

doi:10.1088/978-0-7503-5482-0ch4

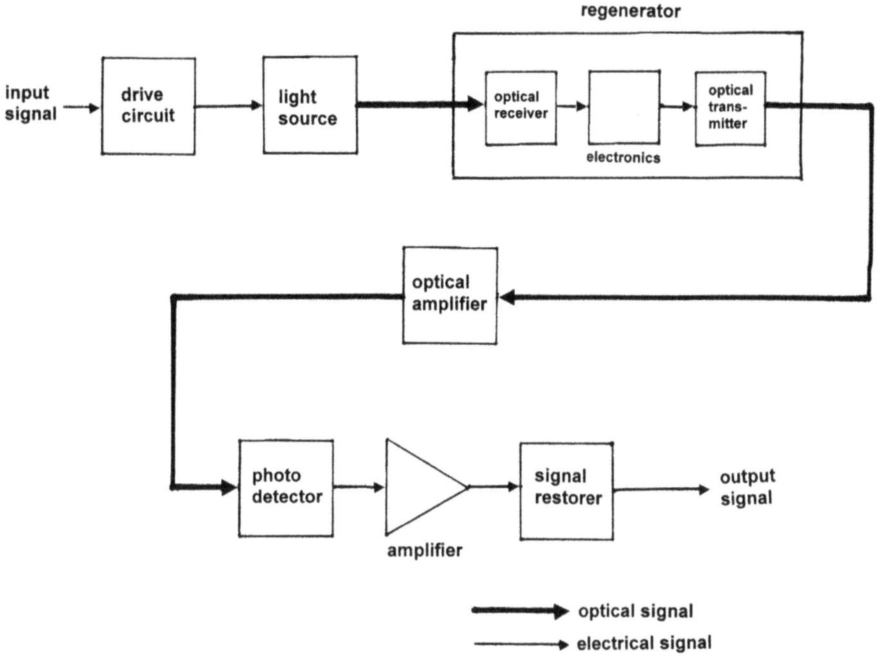

Figure 4.1. A schematic of the general design of a fiber optic communication system.

Fiber optic communications have some advantages over conventional electrical conductors. These include:

- small size and low weight
- the absence of electromagnetic interference
- high security
- low attenuation
- high bandwidth

The absence of electromagnetic interference is due to the fact that the signal is carried by a light beam, rather than as an electric current in a wire. Since fiber optic cables do not produce a surrounding electromagnetic field, it is not possible to detect the signal externally, as in the case of an electrical transmission, leading to increased security of data transmission.

The limits to the distance between transmitter and receiver and to the frequency of the signal are the result of two factors: attenuation and dispersion. Attenuation is caused by light absorption within the cable, while dispersion is due to the properties of the glass itself and the characteristics of the light. It is typically the latter factor which limits the signal-carrying ability of the fiber.

To analyze the signal transmission in an optical fiber, we begin by expressing the speed of light in a medium with an index of refraction n as

$$v_p = \frac{c}{n},$$

where c is the speed of light in vacuum and v_p is the phase velocity of the wave in the medium, i.e. the speed of the wavefronts. The group velocity of the wave, v_g, is the speed of the signal pulses, i.e. the speed at which information can be transmitted. This velocity is

$$v_g = c\left(n - \lambda\frac{dn}{d\lambda}\right)^{-1},$$

where λ is the wavelength of the light source. Since the light output from an LED or laser diode is not truly monochromatic but has some spectral bandwidth, it is important to understand the relationship between the index of refraction and the wavelength. Figure 4.2 shows the index of refraction as a function of wavelength for some common optical glasses. Since the index of refraction is a function of wavelength, the group velocity of the signal is also a function of wavelength. This property means that if we send a pulse of light through an optical fiber, then the shorter-wavelength component of the light travels slower than the longer-wavelength component of the light, and the pulse becomes stretched out, both spatially and temporally. The farther the pulse travels in the fiber, the more it is stretched out. Therefore, if we modulate a light beam in order to transmit a signal over a fiber optic cable, then the maximum frequency of modulation is limited by the length of the cable, the dispersion of the glass fiber, and the spectral width of the light source.

From a practical standpoint, the length of the cable is determined by the spacing of regenerators, as shown in figure 4.1. This distance might be a kilometer or so.

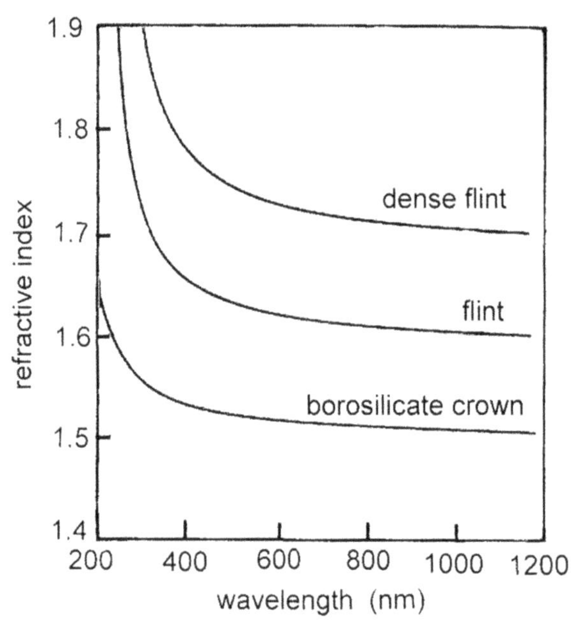

Figure 4.2. Dispersion relations for some common types of glass. The range of visible wavelengths is from about 390 nm to about 700 nm.

On the basis of figure 4.2, it is preferable to use light that has as long a wavelength as possible, as the dispersion, $dn/d\lambda$, decreases with increasing wavelength. Perhaps the most important consideration, though, is the spectrum of the light source. If an LED is used, then figure 3.10 shows that the light will have a spectral width of about 35 nm. This spectral width translates into a maximum data transmission rate of about 500 MHz. Laser diodes can have spectral widths in the order of 2 nm. The corresponding maximum data transmission rate is about 10 GHz. This simple example clearly illustrates the advantages of the greater monochromaticity of laser radiation for data communication applications.

4.3 Optical data disks

Optical data disks are in common use in the form of CDs, DVDs, and Blu-ray disks (for example) for the storage of information for computers and audio and video devices. There are many aspects to the operation and use of optical data disks, including the method of data encryption, the computer interface, the rotation mechanism, and the head-positioning mechanism. Here, we concentrate on the optical system and the application of laser diodes for writing and reading data on the disc.

Three different designs of optical media are used for data storage; read-only media (ROM), recordable media, and rewritable media. ROM are produced by a manufacturing process that creates pits on the surface of the disc. These pits are arranged in spiral grooves on the surface of the disc, much like the grooves on the surface of a vinyl phonograph record, although on a much smaller scale. The depth of the pits is typically of the order of about 20% of the wavelength of the laser light. The reflected beam is phase shifted relative to the incident beam and interferes in a manner that depends on the depth of the pit. Thus, data is encoded on the disc by manufacturing pits of varying depth along the spiral groove on the disc. A simple optical arrangement for reading data from such a disk utilizes a photodiode that detects the reflected light. The operation of ROM optical drives depends on the intensity and coherence of the laser radiation, which allows the light to be focused down onto a very small spot in order to achieve high information density on the disc. The monochromaticity of the radiation allows well-defined interference effects that depend on the depth of the pits on the disc surface.

Information is 'burned' onto the surfaces of recordable disks using light from a laser diode to selectively heat small regions of the disc. These disks are coated with an organic dye which changes its optical reflectivity when heated. The reflectivity is controlled by changing the laser output power. When the disc is read, regions of higher or lower reflectivity are analogous to the pits of varying depth on a ROM disc. Higher write speeds need faster heating rates and therefore require more powerful laser diodes. The lasers used to write information to recordable media might have an output of 200 mW, compared to the lasers of about 5 mW used for reading ROM disks.

While the changes to the reflectivity of the organic dye introduced by heating the surface of a recordable disc are permanent, rewritable disks use a different mechanism that is reversible. These disks are coated with a crystalline alloy. Heating this alloy with a laser allows the alloy to be melted and re-solidified in an amorphous phase. Since the different phases have different reflectivities, information can be stored by controlling phase formation through the laser output. Since the alloy can be remelted by reheating, new data can overwrite old data.

4.4 Printers

The general design of a laser printer is illustrated in figure 4.3. The printer produces an image on a sheet of paper using a seven-step process, as follows:

1. Raster image processing
2. Charging
3. Exposing
4. Developing
5. Transferring
6. Fusing
7. Cleaning

Each of these steps is described in some detail below.

4.4.1 Raster image processing

A computer converts the object to be printed into a bitmap in which each horizontal line or raster line of the image is represented by a series of bits. The computer sends each raster line to the laser printer in succession without pauses (as occur in the case of inkjet printers).

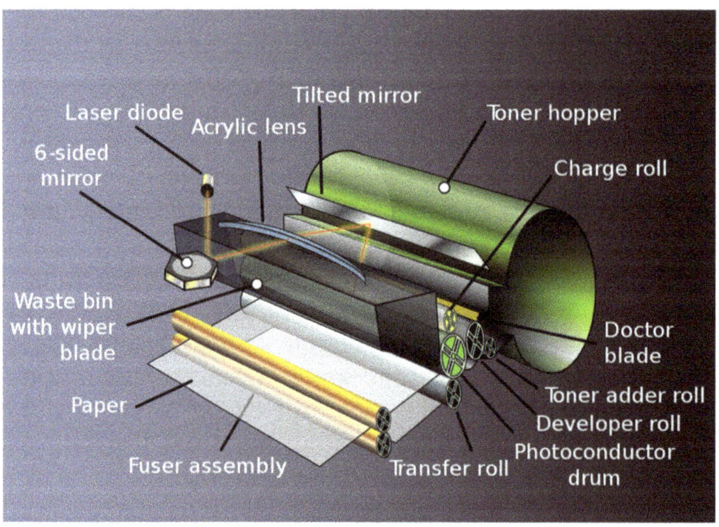

Figure 4.3. The general design of a laser printer. CC BY-SA 4.0. This [Diagram of a cartridge from a laser printer] image has been obtained by the author from the Wikimedia website where it was made available under a CC BY-SA 4.0 licence. It is included within this book on that basis. It is attributed to KDS4444.

4.4.2 Charging

The image to be printed is transferred to a cylindrical printer drum which is made of a photosensitive material. This material is made of layers of semiconducting material, generally organic semiconductors, hence the usual name organic photo conductor (OPC). Before the image is transferred to the printer drum, a negative charge (electrons) is deposited on the surface of the drum. This charge is deposited by a wire (called the corona wire) which is at a high negative potential and is placed near the surface of the drum. As the drum rotates, electrons collect on the surface and form a uniform layer of charge on the surface of the drum. As the OPC is relatively nonconductive, the charge remains on the surface at this point.

4.4.3 Exposing

The image is transferred to the drum by exposing selected portions of the drum to a laser beam. The laser is modulated by the information stored in the raster file and is scanned across the drum as the drum rotates. Areas where the image is intended to be dark are exposed by the laser, but the laser is turned off while scanning regions which will be white in the final print. When the laser beam strikes the surface of the photosensitive drum, the light is absorbed by the photosensitive surface of the drum. This creates charge carriers which increase the conductivity of the drum and allow the charge on the drum to be carried away in these regions. Thus, the image to be printed becomes a pattern of uncharged dots on the surface of the drum.

4.4.4 Developing

At this point, the surface of the drum is exposed to the toner particles. The toner consists of fine particles of a coloring agent (e.g. carbon black for a black and white printer) mixed with powdered plastic. These toner particles are given a negative charge and are preferentially deposited on the regions of the drum which have been exposed by the laser and have lost their negative charge. The toner particles are repelled from the regions which have not been exposed and which still retain their negative charge. Thus, the image to be printed is represented by dark toner particles on the surface of the drum.

4.4.5 Transferring

A sheet of paper is then placed in contact with the drum; as the drum rotates, the paper is moved along with the drum and the toner particles are transferred from the drum to the paper. Some printers use a positively charged electrode on the other side of the paper to help pull the negatively charged toner particles off the drum.

4.4.6 Fusing

The paper containing the image is then passed between two rollers and heated. The combination of heat and pressure melts the plastic component of the toner particles and fuses the image onto the paper.

4.4.7 Cleaning

The final step of the printing process is to remove residual toner and charge from the drum so that it is ready to print the next page. The remaining charge is removed by exposure to a discharge lamp and the unused toner is scraped off by a soft plastic blade.

4.5 Industrial applications

The industrial use of lasers most commonly takes advantage of their high intensity and their ability to be focused onto a very small area to heat or to melt material. Their principal applications include cutting, welding, selective heat treatment, and additive manufacturing. For most applications, CO_2 gas lasers or Nd-based (e.g. Nd-YAG (yttrium aluminum garnet)) solid-state lasers are used, as these produce high outputs and are reasonably cost-effective. For cutting applications, the laser beam is focused onto a small spot on the material (as illustrated in figure 4.4) and a gas jet is used to carry away excess material. Computer-controlled laser movement allows complex patterns to be cut.

For welding applications, the ability to use CO_2 or Nd lasers in either the continuous mode or the pulsed mode allows for versatility in positioning welds at the desired locations. The high heating rates that can be achieved in laser welding are beneficial in the welding of dissimilar metals with greatly different thermal conductivities (e.g. copper and stainless steel), which is difficult with conventional welding techniques.

The traditional manufacturing methods used for the production of metal components include machining (i.e. the removal of material to achieve the desired shape), casting of molten material, and powder sintering in a mold. These methods

Figure 4.4. Laser cutting of a metal sheet. Guryanov Andrey (2023) / Shutterstock.com.

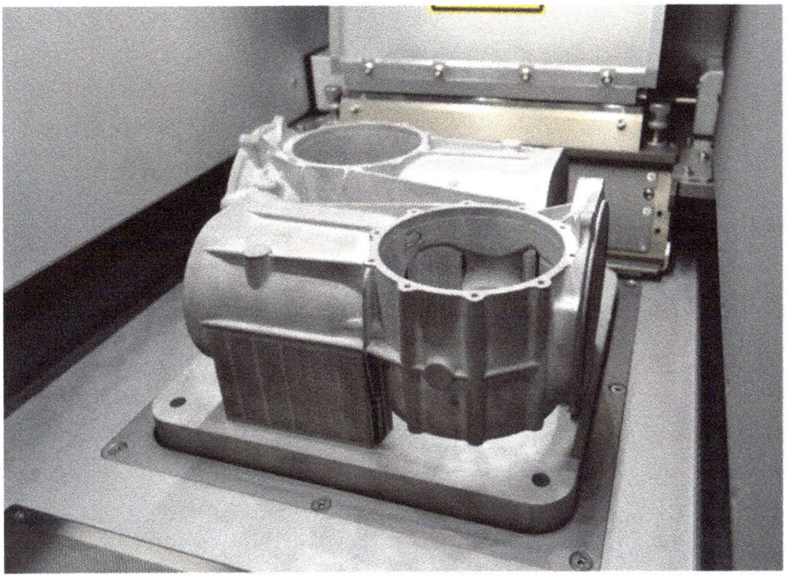

Figure 4.5. A component created by the additive manufacturing processes. MarinaGrigorivna (2023) / Shutterstock.com.

have geometric limitations, which prohibit the production of certain shapes, e.g. a single-piece hollow sphere. The production of complex shapes can also be time-consuming and expensive. Additive manufacturing is an alternative to these methods along the lines of 3D printing. In additive manufacturing, a thin layer of powder is deposited on a build table and a laser is rastered across the surface by a computer-controlled optics system to form the cross section of the object. The build platform is then lowered, and a new layer of powder is applied and laser melted. This approach is an effective and economical way to produce complex components or individual components for prototypes. Figure 4.5 shows a component manufactured using this technique that would be difficult or impossible to produce as a single piece by conventional methods.

4.6 Photolithography

The term photolithography refers to the process of using light to produce a patterned thin film on a substrate. It is the most commonly used method of producing appropriate patterns on a semiconducting wafer (usually silicon) for the purpose of manufacturing an integrated circuit. The pattern may be produced either by projecting an image onto the surface or by exposing a suitable mask in direct contact with the surface to a source of light. The use of a sequence of masks on the surface of the semiconducting substrate allows for the production of circuit components by subsequent etching, deposition, or implantation processes. In recent years, excimer lasers (see section 2.6) have become the standard light source for photolithography. This is because their high output power allows for short exposure times and their short wavelength allows for the creation of patterns with extremely small features.

The photolithographic process used for the preparation of silicon-based semiconductor devices consists of several steps (as illustrated in figure 4.6 and outlined below).

4.6.1 Preparation of the wafer

The silicon wafer is first cleaned to remove any surface contamination. Hydrogen peroxide, acetone, and methanol are cleaning solutions commonly used for this purpose. The wafer is then heated in an oxygen atmosphere to create a layer of insulating silicon dioxide on the surface.

4.6.2 Application of photoresist

The photoresist is a light-sensitive organic material that is used to produce the desired pattern on the surface of the wafer. In general, there are two types of photoresist. Positive photoresist, as illustrated in the example in figure 4.6, is a material that becomes more soluble in certain solvents when exposed to the appropriate radiation (e.g. ultraviolet light (UV)) and can therefore be readily removed. Negative photoresists become hardened upon exposure and become less soluble. The wafer is coated with photoresist dissolved in a suitable solvent by spin coating, which produces an even layer of the photoresist. The final thickness of the photoresist is an important factor in the ability of the photolithographic process to produce small details in the resulting semiconducting device. The 'islands' of photoresist can become unstable if their aspect ratio (height to width ratio) is too large. Typically, aspect ratios of 4:1 are acceptable. After the wafer is coated with the photoresist, it is heated to drive off the photoresist solvent.

4.6.3 Alignment of the photomask

The mask containing the geometric details of the circuit components is then applied and aligned. The details of the alignment procedure are somewhat different for contact masks (as shown in the figure) and projected masks.

4.6.4 Exposure to UV light

After the mask is applied, the wafer is exposed to intense UV radiation. The photoresist undergoes changes as described above for positive or negative photoresists. This allows the selected regions of the photoresist to be removed as described below.

4.6.5 Development and removal of photoresist exposed to UV light

After exposure, the photoresist is developed. The developer is a chemical that removes the relevant portion of the photoresist. In the case of a positive photoresist, the portion of the photoresist that has been exposed to UV radiation becomes soluble in the developer and is removed from the surface of the wafer. The developer

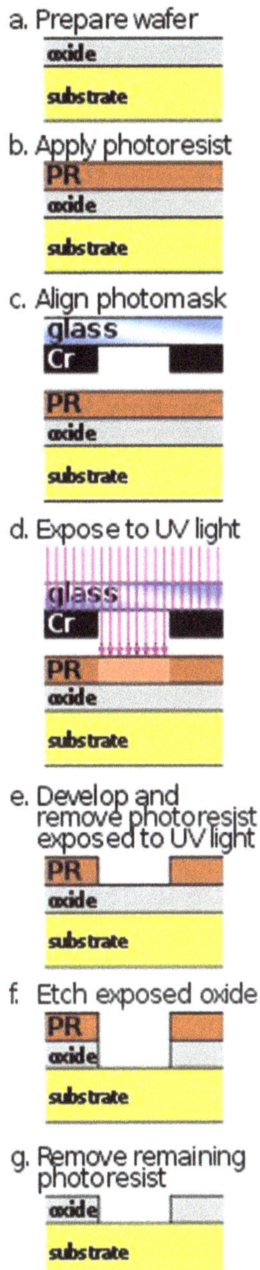

Figure 4.6. A photolithographic process that uses an ultraviolet laser to produce integrated circuit components, as described in the text. This [Simplified illustration of dry etching using positive photoresist during a photolithography process in semiconductor microfabrication] image has been obtained by the author from the Wikimedia website where it was made available under a CC BY-SA 3.0 licence. It is included within this book on that basis. It is attributed to Cmglee.

is typically evenly distributed on the surface of the wafer using a spinner. Metal-free developers are most commonly used to avoid the possibility of metal ions affecting the conductive properties of the integrated circuit components. Heating the wafer after the development stabilizes the remaining photoresist.

4.6.6 Etching of exposed oxide

Etching may be used to remove layers of material that have been deposited on the substrate and which have been exposed in areas where the photoresist has been removed by the developer following the exposure. In the example in figure 4.6, the insulating oxide layer is removed. The etching may be accomplished by one of two methods: wet etching, which typically uses an acid to remove material, or dry etching, which uses a plasma for the same purpose. In either case, the chemical composition of the etchant is chosen to be compatible with the nature of the material to be removed.

4.6.7 Removal of remaining photoresist

After the etching process, the remaining photoresist which protected the underlying layer during etching must be removed. This may be done either with a suitable chemical solvent or by an oxygen-containing plasma which oxidizes the remaining photoresist.

The photolithographic process can be repeated (in some cases up to 50 times) to create the layers of semiconductors, insulators, and conductors necessary for an integrated circuit. Regions of exposed semiconductor created during this process can be implanted with impurities to produce the required semiconducting junction devices in the circuit.

The excimer lasers most commonly used for this purpose are the ArF laser (which has a wavelength of 193 nm) and the KrF laser (which has a wavelength of 248 nm). As noted above, the short wavelengths of these ultraviolet lasers are advantageous for the production of circuits with a high integration density.

4.7 Inertial confinement fusion

Inertial confinement fusion (often referred to as laser fusion) is one of the approaches for producing energy by fusing light nuclei together to make a heavier nucleus. Energy is liberated because of the different binding energies before and after the fusion. Fusion is the method by which energy is produced in the Sun (and all other stars). Most of the energy produced by the Sun comes from the fusion of four hydrogen nuclei into one helium nucleus. This process is not straightforward and proceeds in several steps. The nucleus of the ^{1}H atom is a proton and the nucleus of the ^{4}He atom consists of two protons and two neutrons. (The superscript before the element name gives the total number of nucleons, i.e. protons plus neutrons, in the nucleus.) So, the process in the Sun not only binds nuclei together but also converts two protons into two neutrons in the process. The most common method of energy production in the Sun begins by fusing together two protons:

$$p + p \rightarrow d + e^{+} + \nu_e, \tag{4.1}$$

where d is a deuteron (that is, the nucleus of a deuterium or ^{2}H atom), e^{+} is the positron or antielectron, and ν_e is the electron neutrino. The process in equation (4.1) is referred to as the p–p process and represents the simultaneous binding together of two protons and the conversion of one of the protons into a neutron, a positron, and a neutrino. This latter process requires the weak interaction and, as a result, proceeds very slowly.

The energy associated with the p–p process can be calculated by taking the difference between the mass of the left-hand side and that of the right-hand side, where this difference in mass, Δm, is converted into energy, E, according to Einstein's mass–energy equivalence relation

$$E = \Delta mc^2,$$

where c is the speed of light. For the p–p process, the energy is, therefore,

$$E = [m_d + m_e - m_p]c^2 ,$$

where the positron has the same mass as that of the electron and the electron neutrino is assumed to be massless. Using known values for the particle masses gives a value for the energy of the p–p process of 0.42 MeV. Normally, the positron would then annihilate an electron in the environment, (usually) giving rise to two gamma rays, γ,

$$e^{+} + e^{-} \rightarrow 2\gamma$$

and yielding an additional $2m_e c^2 = 1.022$ MeV of energy, for a total of 1.44 MeV. There are a number of possible fusion processes involving deuterons, but because the Sun still largely comprises ^{1}H, the most likely process is the fusion of a deuteron and a proton,

$$d + p \rightarrow {}^{3}He + \gamma, \tag{4.2}$$

which releases 5.49 MeV of energy. The most likely route from here is the eventual fusion of two ^{3}He nuclei,

$${}^{3}He + {}^{3}He \rightarrow {}^{4}He + 2{}^{1}H + \gamma, \tag{4.3}$$

which yields 12.86 MeV of energy.

A complete fusion cycle, therefore, requires the process in equation (4.1) to occur twice, the process in equation (4.2) to occur twice, and the process in equation (4.3) to occur once. The total input to the fusion process is six protons, and the final output is a ^{4}He nucleus, two protons, two positrons, and two neutrinos, leading to the net reaction

$$4p \rightarrow {}^{4}He + 2e^{+} + 2\nu_e.$$

The total energy, including the electron–positron annihilations is $2 \times 1.44 + 2 \times 5.49 + 12.86 = 26.72$ MeV. The magnitude of this number can be appreciated if the energy is converted to Joules per kg of hydrogen; 6.5×10^{14} J kg^{-1}. This may be compared to the energy produced in the chemical reaction of burning oil, 3.8×10^{7} J kg^{-1}, which clearly justifies the interest in developing this source of energy, as substantial amounts of energy can be produced from an inexpensive fuel (e.g. the protons of hydrogen atoms in water).

One might think that if a reactor could be constructed that reproduced the conditions in the center of the Sun, then this would represent a viable source of energy. However, this is not true. The Sun will require in the order of 20 billion years to fuse all its hydrogen into helium (and subsequently into heavier elements). If we put fuel into a reactor, then we obviously do not want to have to wait tens of billions of years to get the energy out of that fuel. Another way of looking at this is to calculate the power produced per unit volume of the Sun. The total power output of the Sun is 3.8×10^{26} W. Although the Sun's diameter is 1.392×10^{6} km, most of the energy is produced near the center, where the temperature and pressure are the highest. Figure 4.7 shows the power produced per m^3 as a function of distance from the center of the Sun. This figure shows that the reactions at the center of the Sun produce as much power per cubic meter as a few medium-sized light bulbs. The only reason that the Sun produces so much power is because it is so large. The problem with implementing the reactions that occur in the Sun in a power plant on Earth is that these reactions require the weak interaction and therefore proceed very slowly.

A reasonable approach to constructing an operational fusion reactor would be to bypass equation (4.1) and start with deuterons. The most obvious reaction would be

$$d + d \rightarrow {}^{4}He + \gamma,$$

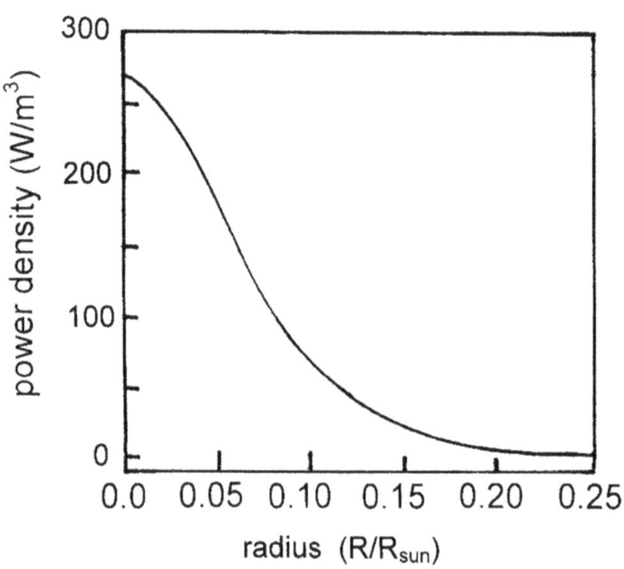

Figure 4.7. Energy production (W m^{-3}) as a function of the distance from the center of the Sun as calculated from the standard solar model. Data adapted from Bahcall (2004).

which releases 23.85 MeV of energy. Because of the large amount of energy released in this single reaction, the ^{4}He nucleus that is formed is unstable and immediately leads to one of the following reactions

$$\mathrm{d} + \mathrm{d} \rightarrow {}^3\mathrm{He} + \mathrm{n} \tag{4.4}$$

or

$$\mathrm{d} + \mathrm{d} \rightarrow \mathrm{t} + \mathrm{p}. \tag{4.5}$$

The nucleus of the right-hand side of equation (4.5) is referred to as a triton (t); it is the nucleus of the ^{3}H atom (tritium). The reactions in equations (4.4) and (4.5) yield 3.27 and 4.03 MeV, respectively, and may be followed by the reaction in equation (4.3) or by

$$\mathrm{d} + \mathrm{t} \rightarrow {}^4\mathrm{He} + \mathrm{n},$$

which yields 17.59 MeV of energy. In all cases, the fusion of deuterium leads to ^{4}He and 23.85 MeV of energy. Although deuterium accounts for only 0.015% of the naturally occurring hydrogen on Earth, there is enough deuterium in only 0.01% of the Earth's seawater to provide all of our energy for about four million years at our current rate of use.

Although a d–d fusion reactor would provide a virtually inexhaustible supply of energy for society, the problem is to create conditions amenable to d–d fusion. The first difficulty we need to overcome is the coulombic repulsion between the two

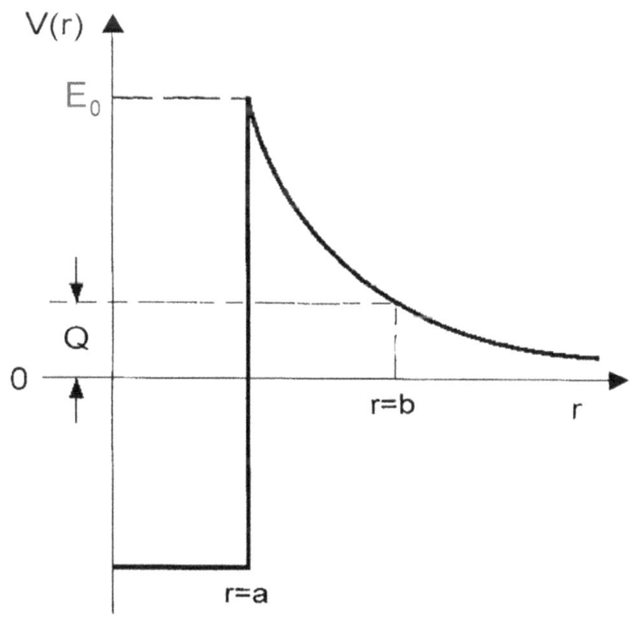

Figure 4.8. The potential between two positively charged nuclei as a function of the distance between their centers.

positively charged nuclei. Figure 4.8 shows the coulombic potential between two nuclei as a function of the distance between their centers. When the nuclei are far apart, they are repelled by the coulombic interaction. If the nuclei approach one another with some kinetic energy, Q, as shown in the figure, then they scatter as a result of the coulombic interaction when they are a distance b apart. If the nuclei actually came into contact at a distance a (in the figure), then the attractive strong interaction between the nuclei takes over and they fuse together. The energy required to do this, E_0 in the figure, is greater than that which can be achieved by any practical means. The philosophy of constructing a fusion reactor is to create an environment in which the nuclei are as dense as possible and have as great an energy as possible, so that a reasonable number of nuclei can quantum mechanically tunnel through the coulomb barrier and fuse together to produce energy. Since the temperatures associated with such energies are far above the melting points of any physical materials, the design of a fusion reactor requires a suitable mechanism for confining the collection of nuclei. At these temperatures, all atoms would be fully ionized, and the material would be a plasma consisting of positively charged nuclei and negatively charged electrons. One approach to confining the plasma is by means of magnetic fields which allow the motion of the charged particles to be controlled. Another approach makes use of lasers in a technique referred to as inertial confinement. Before we continue with a description of inertial confinement, let us look at the actual conditions that are necessary to achieve a fusion reaction.

The probability of fusion (i.e. the probability for nuclei to tunnel through the coulomb barrier) increases with increasing temperature (i.e. the kinetic energy of the nuclei). Figure 4.9 shows the reactivities of d–d, d–t, and d–^{3}He fusion as a function of temperature. The reactivity is defined as the average of the product of the fusion

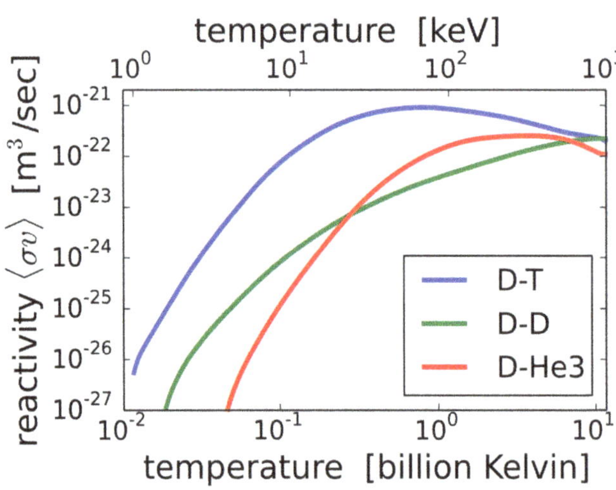

Figure 4.9. Reactivity as a function of temperature for d–d, d–t, and d–^{3}He fusion. CC BY 2.5 by Dstrozzi. This Dstrozzi (2009) image has been obtained by the author from the Wikimedia website where it was made available under a CC BY 2.5 licence. It is included within this book on that basis. It is attributed to Dstrozzi.

cross section, σ, and the relative velocity of the two nuclei, v, that is, $\langle \sigma v \rangle$. The fusion rate, f, is given in terms of the reactivity by

$$f = n_1 n_2 \langle \sigma v \rangle,$$

where n_1 and n_2 are the number densities of the two nuclear species.

It is clear from figure 4.9 that the fusion rate of d–t fusion is much higher (at any given temperature) than the fusion rate of d–d fusion. Thus, much lower temperatures, by a factor of up to ten or even more, are needed to achieve the same fusion rate for d–t fusion compared to d–d fusion. It is for this reason that virtually all current fusion research aims to construct a viable power reactor based on d–t fusion.

Once the appropriate conditions have been achieved, it is anticipated that nuclei may fuse by tunneling through the coulomb barrier at a sufficient rate to yield a net power output. The longer the time that the appropriate conditions are maintained, the greater the probability that fusion will actually occur. The relevant parameter that determines the fusion rate is the Lawson parameter. This parameter is defined as the product of the number density of particles and the confinement time, $n\tau$. For d–d fusion, the Lawson criterion must satisfy the condition $n\tau > 5 \times 10^{21}$ s m^{-3} in order to produce a net energy output. For d–t fusion, the condition is $n\tau > 2 \times 10^{20}$ s m^{-3}. Again, this emphasizes the fact that d–d fusion is more difficult to achieve than d–t fusion.

The objective of laser fusion experiments is to use a high-power laser to heat a very small pellet containing a mixture of deuterium and tritium to a very high temperature. In the process, the fuel is also compressed to a very high density and contained by inertial forces, hence the terminology 'inertial confinement fusion.' Inertial confinement fusion systems as classified as either 'direct drive' or 'indirect drive.' In the former case, the laser radiation is directly incident on a millimeter-sized pellet of fuel. In the latter case, the fuel pellet is contained inside a centimeter-sized cylinder called a hohlraum which is typically made of gold, as shown in figure 4.10. When the laser radiation strikes the hohlraum, the gold is rapidly heated to a very high temperature and emits x-rays. The x-rays are produced by two processes: first, blackbody radiation which, because of the high temperatures involved, is in the x-ray region of the electromagnetic spectrum; second, as a result of induced electronic transitions. Figure 4.11 shows the typical spectrum of a hohlraum heated by laser irradiation, illustrating the broadband blackbody radiation and the series of sharp lines produced by discrete electronic transitions. A major factor favoring indirect drive for laser fusion experiments is related to beam uniformity, and the most successful laser fusion experiments to date fall into this category.

When the laser or x-ray radiation is incident on the fuel, the fuel is heated and compressed in a series of steps as shown in figure 4.12. The incident energy is absorbed by the fuel pellet, which is heated from the outside. This causes the outer portion of the pellet to expand, thereby compressing the inner portion of the pellet through inertial forces. Temperatures in the range of tens of millions of kelvins and densities of up to 10^5 kg m^{-3} can be achieved using this approach.

The most successful laser fusion experiments have been performed at the National Ignition Facility located at the Lawrence Livermore National Laboratory in

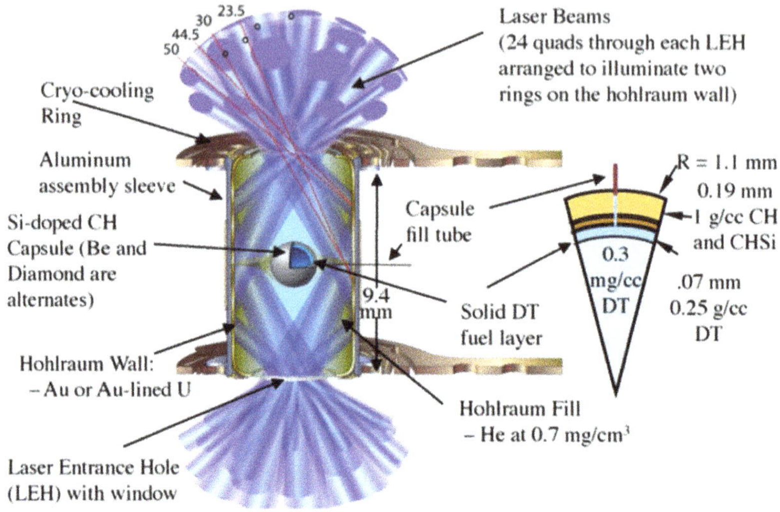

Figure 4.10. (left) The indirect drive scheme used in laser fusion experiments, showing the hohlraum and (right) the internal structure of the fuel capsule. CH = carbon–hydrogen plastic ablator, DT = deuterium–tritium fuel. Reprinted from Landen *et al* (2012) © IOP Publishing. Reproduced with permission. All rights reserved.

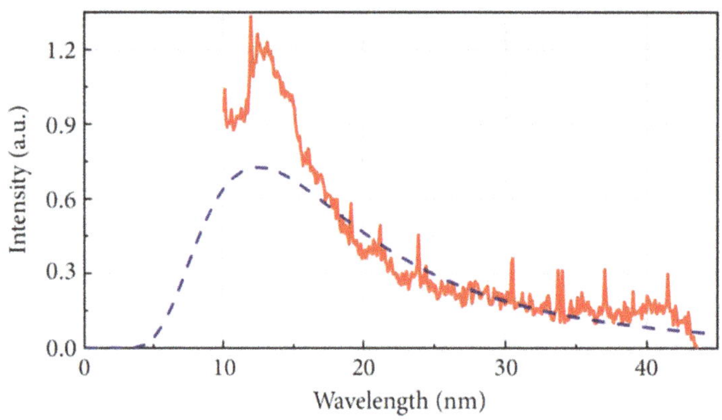

Figure 4.11. The spectrum of a laser-irradiated gold hohlraum. The red line is the measured spectrum and the broken black line is a 20 eV blackbody curve. Reprinted from Ma *et al* (2022). CC BY 4.0. Copyright © 2022 The Authors.

California; see figure 4.13. This facility utilizes a Nd-glass laser with pumping radiation provided by 7680 xenon flash lamps and a lasing transition at 1053 nm (in the infrared). The output of this laser is more than 5×10^{14} W. This may be compared with the average total global power consumption of about 1.5×10^{13} W. The laser is pulsed with a pulse duration of about 4 ns, giving an energy per pulse of $E = Pt = (5 \times 10^{14}\ \text{W}) \times (4 \times 10^{-9}\ \text{s}) = 2$ MJ. Although this is about the same as the

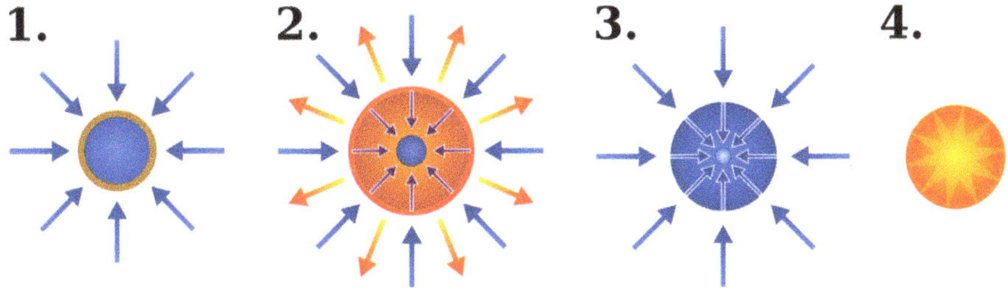

Figure 4.12. The sequence of events in the irradiation of a d–t fuel pellet: (1) a laser beam or laser-produced x-rays are incident on the fuel pellet; (2) the absorption of x-ray radiation and ablation of the outer portion of pellet, which create an inward force on the inner portion of the pellet; (3) the compaction of the inner portion of the pellet; and (4) the creation of thermonuclear fusion. This [The stages of inertial confinement fusion] image has been obtained by the author from the Wikimedia website, where it is stated to have been released into the public domain. It is included within this book on that basis. It is attributed to B D Esham.

Figure 4.13. One of the two laser bays at the National Ignition Facility, Lawrence Livermore National Laboratory. The scale can be seen from the workers at the lower left of the image. Courtesy of Lawrence Livermore National Laboratory.

energy utilized by an automobile while traveling half a kilometer, it occurs within a few nanoseconds and is concentrated on a target weighing a few milligrams.

The absorption of laser radiation by the fuel at 1053 nm is relatively poor, but it is much greater at shorter wavelengths. Harmonic generation, as described previously, is useful for producing photons of shorter wavelength and greater energy. In the case of the National Ignition Facility, a combination of two potassium dihydrogen phosphate crystals is used to generate third-harmonic radiation at 351 nm. This radiation is in the ultraviolet region, where the power absorption efficiency is near 100%. This laser radiation is incident on a hohlraum, as shown in figure 4.10, which then produces x-rays that irradiate the fusion fuel.

The National Ignition Facility was completed in 2010 (see Dunlap 2021) and has recently announced a major breakthrough in inertial confinement fusion research. On 5 December 2021, the National Ignition Facility achieved a fusion energy output of 3.15 MJ for a laser energy input into the holhraum of 2.05 MJ (Kramer 2023). While it is very significant that the fusion energy has exceeded the laser energy, this falls short of a viable energy source in which the fusion energy must exceed total the energy consumption. In particular, the concerns that need to be addressed before laser fusion could become a viable source of energy include:

- the low efficiency of Nd-glass lasers (typically 1.5% for the conversion of electrical energy to laser energy)
- the need for frequency tripling due to the long wavelength of the lasing transition
- the low repetition rate (typically a few shots per day) due to overheating of the lasing medium
- the low laser beam uniformity, which requires the use of indirect drive

Many recent laser fusion experiments have, to a large extent, focused on overcoming the drawbacks of the Nd-glass laser approach used at the National Ignition Facility. Two laser technologies that have received some attention in recent years in this regard are excimer lasers (see section 2.6) and diode-pumped solid-state (DPSS) lasers (see section 3.4).

Excimer lasers have been utilized for applications such as photolithography, as noted previously, because of their high output power. Excimer lasers and KrF lasers in particular have been considered as possible drivers for inertial confinement fusion reactors. The advantages of KrF lasers over Nd-glass lasers for this application include:

- high efficiencies (up to about 10%) in the conversion of electrical energy to laser radiation
- a short wavelength (248 nm) that does not require harmonic generation
- high repetition rates (up to about 5 Hz for extended periods of time) as a result of the continuous flow of the lasing medium through the laser channel
- high beam uniformity, which allows the use of the direct drive approach

The use of excimer lasers for inertial confinement fusion has been considered most seriously by the U.S. Naval Research Laboratory in Washington, DC, which has constructed two prototype systems that have clearly demonstrated the advantages of excimer lasers for fusion experiments. Unfortunately, at present, excimer lasers are limited to output energies of around 3 kJ per pulse. Although this is sufficient for many other laser applications, it falls short of the requirements for successful laser fusion by nearly three orders of magnitude.

Diode-pumped solid-state (DPSS) lasers constitute another possible system that could overcome the deficiencies of traditional flash-tube-pumped Nd-glass lasers. The high efficiency of DPSS lasers (up to 20%) allows for high repetition rates due to minimal heating of the lasing medium. The experimental DPSS Mercury laser system at Lawrence Livermore National Laboratory uses Nd-doped ytterbium–strontium flouroapatite crystals and has operated for extended periods of time at a repetition rate of 10 Hz. At present, the maximum energy per pulse is about 50 J.

While laser fusion experiments have not produced net energy, they have provided some interesting results. They have identified the laser features that require improvement and have shown which laser designs satisfy particular fusion requirements. Ongoing research in this field will hopefully yield progress toward a viable design.

The above examples are a few of the very large number of applications that lasers have found. Other notable applications include medical diagnostics and treatment, in which (for example) the tunability of dye lasers allows for the optimization of the interaction between the laser radiation and particular tissues. Laser pointers and barcode readers utilize laser diodes and are in common use worldwide. Holography, which was originally only seen in scientific research laboratories, has found widespread use in security devices that appear on the banknotes of numerous countries and on credit cards.

References and suggestions for further reading

Bahcall J 2004 Standard Solar Model (BP2004) http://www.sns.ias.edu/~jnb/SNdata/Export/BP2004/bp2004stdmodel.dat

Cmglee 2022 Photolithography etching process https://commons.wikimedia.org/wiki/File:Photolithography_etching_process.svg

Dstrozzi 2009 Plot of the fusion reactivity (average of cross-section times relative speed of reacting nuclei) vs. temperature for three common reactions. https://commons.wikimedia.org/wiki/File:Fusion_rxnrate.svg

Duarte F J 2008 *Tunable Laser Applications* 2nd edn (Boca Raton: CRC Press)

Dunlap R A 2004 *An Introduction to the Physics of Nuclei and Particles* (Belmont, CA: Brooks/Cole)

Dunlap R A 2018 *Novel Microstructures for Solids* (San Rafael: Morgan & Claypool)

Dunlap R A 2021 *Energy from Nuclear Fusion* (Bristol: IOP Publishing)

Esham B D 2007 The stages of inertial confinement fusion https://commons.wikimedia.org/wiki/File:Inertial_confinement_fusion.svg

Grigorivna M 2023 A model with supports created in a laser sintering machine stays in the working chamber. https://www.shutterstock.com/image-photo/model-supports-created-laser-sintering-machine-766905421?src=Av3MAYQn6ra3fm2WFSlSRQ-1-62

Guryanov A 2023 Cutting of metal. Sparks fly from laser https://www.shutterstock.com/image-photo/cutting-metal-sparks-fly-laser-318330188?src=X0FfSFdT53guNPwpJvi4Cg-1-14

KDS4444 2016 Diagram of a cartridge from a laser printer https://commons.wikimedia.org/wiki/File:Laser_toner_cartridge.svg

Kramer D 2023 NIF success gives laser fusion energy a shot in the arm *Phys. Today* **76** 25–7

Krane K S 1987 *Introductory Nuclear Physics* (New York: Wiley)

Landen O L *et al* 2012 Progress in the indirect-drive national ignition campaign *Plasma Phys. Control. Fusion* **54** 124026
Ma B *et al* 2022 Plasma spectroscopy on hydrogen–carbon–oxygen foam targets driven by laser-generated hohlraum radiation *Laser Part. Beams* **2022** 3049749
Schneider M B *et al* 2008 An overview of EBIT data needed for experiments on laser-produced plasmas *Can. J. Phys.* **86** 259–66
Wilson J and Hawkes J F B 1987 *Lasers: Principles and Applications* (New York: Prentice-Hall)

Part II

The application of lasers to the cooling and trapping of atoms

Richard A Dunlap

Chapter 5

Laser cooling of atoms

5.1 Introduction

In the first part of this book, the basic operation of different types of laser was presented along with a variety of applications. Most of these applications are of commercial interest and utilize the monochromaticity, as well as the high power density, associated with laser light. The second part of this book deals with a very different approach to utilizing laser radiation, namely, the cooling and trapping of collections of atoms. This technology utilizes the monochromaticity of laser radiation along with the tunability of some types of laser. In this use of laser radiation, the interaction between photons produced by low-power lasers and individual atoms is used to control atomic movements.

In order to utilize lasers to manipulate atoms, the thermal energy of the atoms must already be very low. This means that the collection of atoms must be cooled to a very low temperature using conventional methods and the laser radiation can then be used to further reduce the temperature and manipulate the movement of the atoms. This chapter begins with a description of the conventional approaches used to produce low temperatures. This is followed by a description of how the temperature can be further lowered using laser radiation.

5.2 The dilution refrigerator

The simplest approach to cooling an object is to make use of the properties of the thermodynamic phase diagram. The element which has the lowest boiling point is helium. Naturally occurring helium consists of two isotopes, ^{4}He, which has a nucleus consisting of two neutrons and two protons and is 99.999 86% naturally abundant and ^{3}He, which has a nucleus consisting of one neutron and two protons and is 0.000 14% naturally abundant. Since ^{4}He is by far the most common, we might want to make use of its properties. The phase diagram of ^{4}He is illustrated in

doi:10.1088/978-0-7503-5482-0ch5

figure 5.1. At a pressure of one atmosphere, liquid helium has a temperature of 4.2 K. We can lower the temperature of liquid helium by lowering its vapor pressure using a vacuum pump. As the vapor pressure is lowered, the liquid remains in thermodynamic equilibrium with the vapor and follows the liquid–vapor transition line, thereby lowering the temperature of the liquid. The details of the temperature–vapor pressure relationship, as illustrated in figure 5.2, show that this method can, in

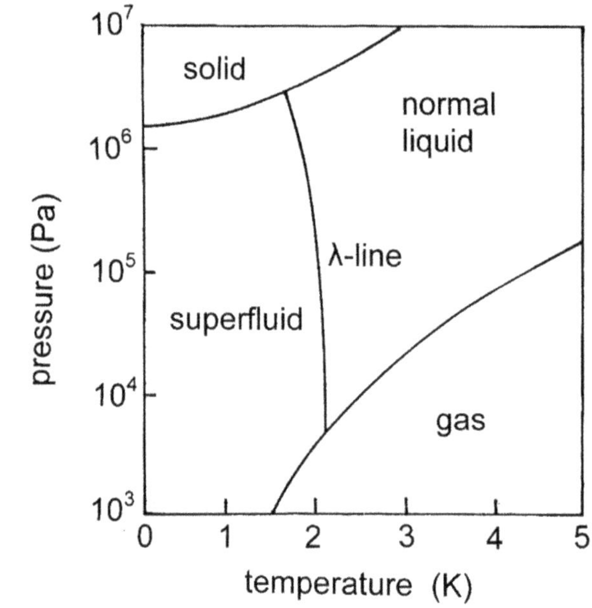

Figure 5.1. The phase diagram of ^{4}He.

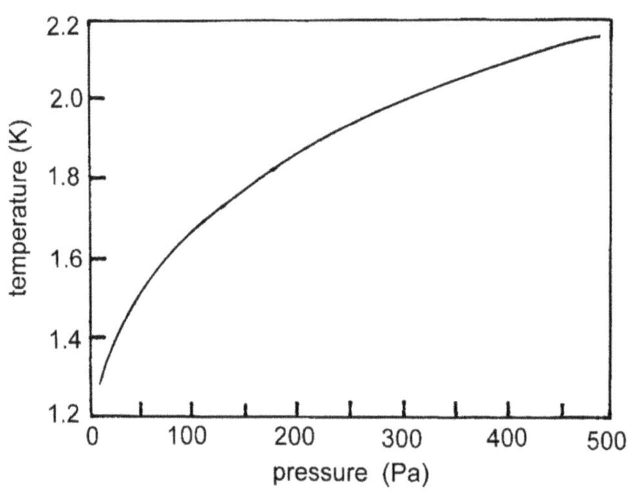

Figure 5.2. The temperature–pressure relationship for ^{4}He.

principle, lower the temperature to about 1.2 K. Lowering the vapor pressure of pure ^{3}He allows the temperature to be reduced to about 0.6 K, a bit lower than for ^{4}He, but still far from the requirements of the experiments described in chapters 8 and 9.

The general technique for achieving temperatures below about 1 K is to use a dilution refrigerator. The dilution refrigerator utilizes the behavior of a mixture of ^{3}He and ^{4}He to transfer heat. The operation of the dilution refrigerator follows from the phase diagram shown in figure 5.3. From about 2.1 K down to 0.87 K, a mixture of ^{3}He and ^{4}He exists in one of two phases, as determined by the temperature and the composition of the mixture. At a given temperature, a mixture of ^{3}He and ^{4}He consists of either ^{3}He in superfluid ^{4}He (for low ^{3}He concentrations) or a mixture of normal ^{3}He and ^{4}He (for higher ^{3}He concentrations). Below the tricritical point at about 0.87 K, a mixture of ^{3}He and ^{4}He phase separates into two components, one with dilute ^{3}He in superfluid ^{4}He (the ^{4}He-rich phase) and one which is normal ^{3}He in a ^{3}He-rich ^{3}He/^{4}He phase. At low temperatures, the ^{4}He-rich phase contains about 6.6% ^{3}He, while the ^{3}He-rich phase approaches pure normal ^{3}He.

The general design of a dilution refrigerator is shown in figure 5.4. The working fluid is ^{3}He (which is analogous to the ammonia in used conventional commercial

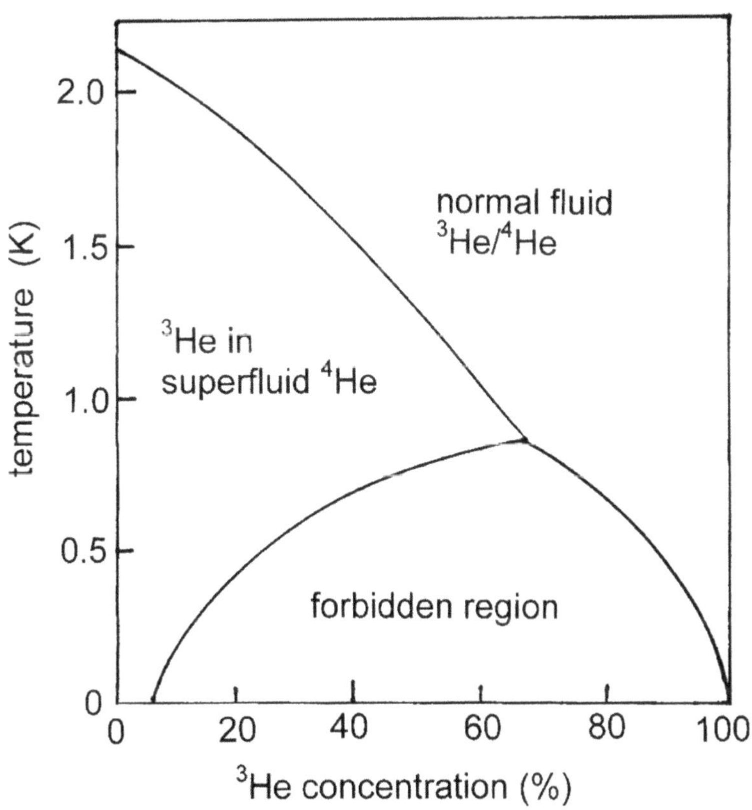

Figure 5.3. The phase diagram of a mixture of ^{3}He and ^{4}He.

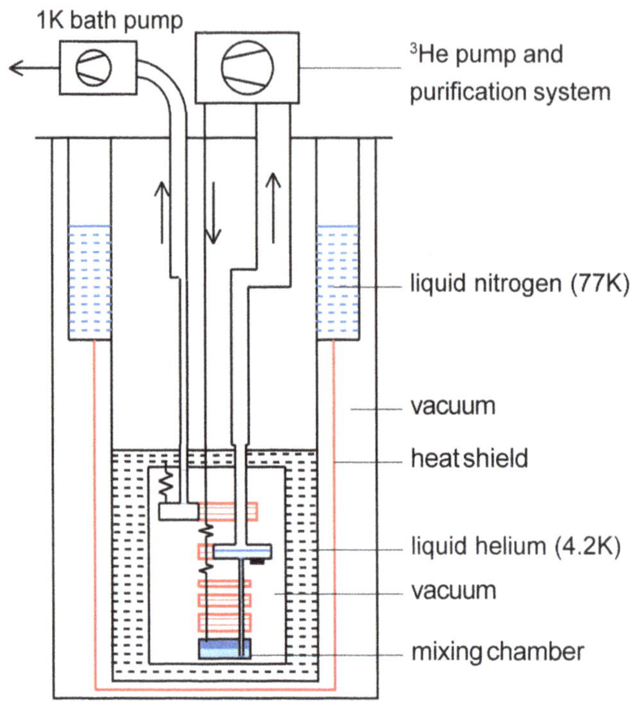

Figure 5.4. A schematic diagram of a ^{3}He–^{4}He dilution refrigerator. This [Schematic diagram of a dilution refrigerator] image has been obtained by the author from the Wikimedia website where it was made available under a CC BY-SA 3.0 licence. It is included within this book on that basis. It is attributed to Adwaele.

refrigerators or the tetrafluoroethane used in household refrigerators), which is circulated through the system by pumps. The principle of operation is described below.

^{4}He at room temperature and a pressure of a few tens of kPa is pumped by the 1 K pot pump (also called the 1 K bath pump) into the refrigerator, as shown in the detail in figure 5.5. The ^{4}He enters into a bath of ^{4}He (the 1 K pot) which is cooled to about 1.2 K by lowering its vapor pressure as described above. This bath condenses ^{3}He gas which enters the condenser (which is inside the 1 K pot), converting it to liquid and removing its latent heat of vaporization. The ^{3}He then passes through the primary impedance (i.e. a constricted region where its pressure is reduced). The ^{3}He is cooled by thermally coupling it through a heat exchanger to the still (operation to be described below). The ^{3}He then enters the mixing chamber, where it is combined with the proper amount of ^{4}He. This ^{3}He/^{4}He mixture phase separates (as it is below the tricritical temperature) into the ^{4}He-rich phase containing about 6.6% ^{3}He and a nearly pure ^{3}He phase. The ^{3}He-rich phase has a lower density than that of the ^{4}He-rich phase and therefore floats on top of it in the mixing chamber. A tube inserted into the mixing chamber connects the region of ^{4}He-rich fluid to the still above it, as shown in the illustration. It is essential that the amount and concentration of ^{3}He in

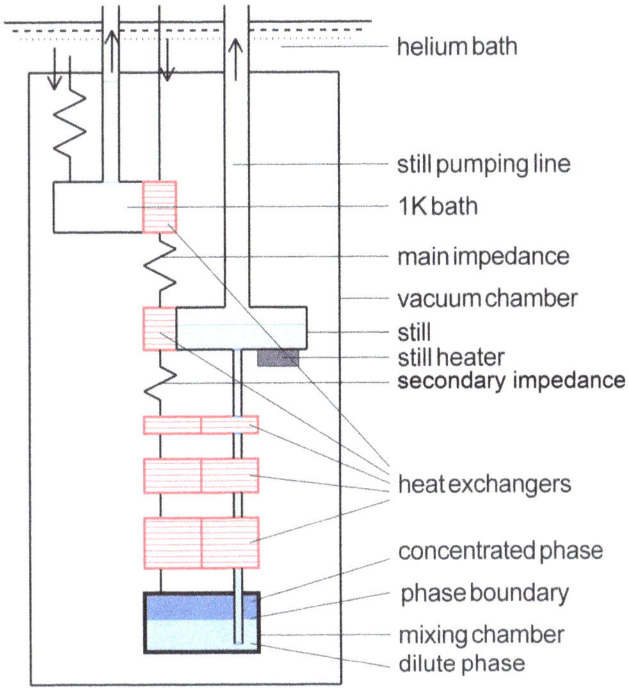

Figure 5.5. Detail of the low-temperature portion of the dilution refrigerator shown in figure 5.4. This [Cold part of dilution refrigerator] image has been obtained by the author from the Wikimedia website where it was made available under a CC BY-SA 3.0 licence. It is included within this book on that basis. It is attributed to Adwaele.

the system is correct so that the interface between the two phases in the mixing chamber and the liquid–gas interface in the still are in the proper locations as shown in the figure.

Cooling to a few mK takes place in the mixing chamber. The ^{3}He concentration gradient across the interface causes ^{3}He in the ^{3}He-rich phase on the top to mix with the ^{3}He dilute phase. As the ^{4}He in the lower part of the mixing chamber does not contribute to the ^{3}He vapor pressure, the dilute ^{3}He in the lower part of the chamber acts like a gas. Thus, the mixing of ^{3}He from the concentrated ^{3}He phase on the top with the dilute ^{3}He on the bottom is equivalent to the evaporation of ^{3}He from the liquid phase, which thereby absorbs the latent heat of vaporization. In this way, the mixing chamber is cooled. In the process, the evaporation of the ^{3}He forces the ^{4}He-rich phase up the tube into the still. Since the concentration of ^{3}He in the ^{4}He-rich phase is finite even as the temperature approaches absolute zero, this method of cooling has, at least in principle, no lower limit, although in actuality a few mK is the practical limit. A pump lowers the vapor pressure in the still, as shown in figure 5.5. Since ^{3}He has a lower boiling point than ^{4}He, it preferentially evaporates in the still. The ^{3}He vapor is now returned to the condenser and the refrigeration cycle repeats. The evaporation of the ^{3}He in the still absorbs heat and is used to cool the ^{3}He as it travels through the impedance stage.

In the next section, we discuss another common method of cooling to a temperature well below the temperature of a pumped helium bath, namely adiabatic demagnetization.

5.3 Adiabatic demagnetization

One of the common methods for lowering the temperature to the range of a few mK is adiabatic demagnetization. Along the lines of the dilution refrigerator, this method makes use of the entropic difference between ordered and disordered phases. In the case of adiabatic demagnetization, the ordered and disordered phases are related to the ordering of magnetic moments in a paramagnetic solid. We begin with a brief overview of some basic magnetic properties of atoms.

The magnetic moment associated with an atom arises primarily as a result of the spins associated with the atomic electrons. This magnetic moment, $\boldsymbol{\mu}$, may be expressed in terms of the spin, orbital, and total angular momentums of the electrons, denoted by S, L, and J, respectively, as

$$\boldsymbol{\mu} = -g\mu_B \boldsymbol{J},$$

where μ_B is the Bohr magneton,

$$\mu_B = \frac{e\hbar}{2m_e}, \tag{5.1}$$

and the g-factor is defined as

$$g = 1 + \frac{J(J+1) + S(S+1) - L(L+1)}{2J(J+1)}.$$

The orientations of the magnetic moments in a material are determined by the interactions between the moments, their interactions with any external magnetic fields, and the temperature. The net magnetization of a sample (per unit volume) is defined as the vector sum of the individual magnetic moments as

$$\boldsymbol{M} = \frac{1}{V}\sum \boldsymbol{\mu}_i,$$

where V is the sample volume. In a paramagnetic material, which is of relevance to magnetic cooling, the interactions between the individual magnetic moments are weak; thus, the net alignment of the magnetic moments and hence the magnetization is determined by the temperature and the external magnetic field defined by the magnetic field strength, $\boldsymbol{H}$. Thus, we may write

$$\boldsymbol{M} = \chi_V(T)\boldsymbol{H},$$

where $\chi_V(T)$ is the temperature-dependent magnetic susceptibility per unit volume. For an ideal paramagnet, the susceptibility is described by Curie's law as

$$\chi_V(T) = \frac{C}{T}. \tag{5.2}$$

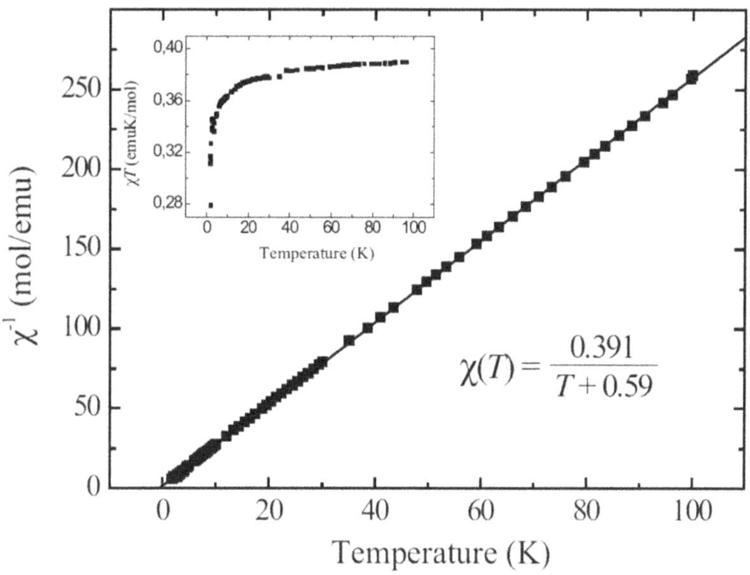

Figure 5.6. The inverse magnetic susceptibility of the copper complex diaqua-(pyridine-2,6-dicarboxylato) copper (II), illustrating the validity of Curie's law in equation (5.2). Reprinted from Pérez *et al* (2017). Copyright 2017 with permission from Elsevier.

The validity of Curie's law is most easily observed by plotting the inverse susceptibility as a function of temperature. This is shown for a paramagnetic material that obeys Curie's law in figure 5.6.

The alignment of the magnetic moments in a paramagnetic material can be used to extract heat from the material. This process, sometimes called adiabatic demagnetization, is more generally referred to as the magnetocaloric effect, as it describes the behavior of materials in which the application of a magnetic field can result in a change in temperature. The change in temperature of a material, referred to as the refrigerant (as it is analogous to the fluid refrigerant used in Carnot cycle refrigerators), is related to the magnetization induced by an applied field H, by

$$\Delta T = -\int_{0_1}^{H} \left[\frac{T}{C(T,H)} \frac{\partial M(T,H)}{\partial T} \right]_H dH,$$

where $C(T,H)$ is the heat capacity of the refrigerant as a function of temperature and applied magnetic field and the derivative is evaluated for a constant field. The change in temperature of the refrigerant may be maximized by optimizing the following experimental parameters:

- using a large applied field
- using a refrigerant with a small heat capacity and
- using a refrigerant that has a large change in magnetization as a function of temperature for a given applied magnetic field.

Figure 5.7 shows the basic design of an adiabatic demagnetization refrigerator. The paramagnetic refrigerant sometimes called a 'salt pill' is shown in the diagram. The operation of the refrigerator is described by the flow diagram in figure 5.8. There are four basic steps in the principle of cooling by adiabatic demagnetization, as follows:

Adiabatic magnetization: A non-magnetized paramagnetic material, the refrigerant, at temperature *T* is thermally isolated from its surroundings. A magnetic field,

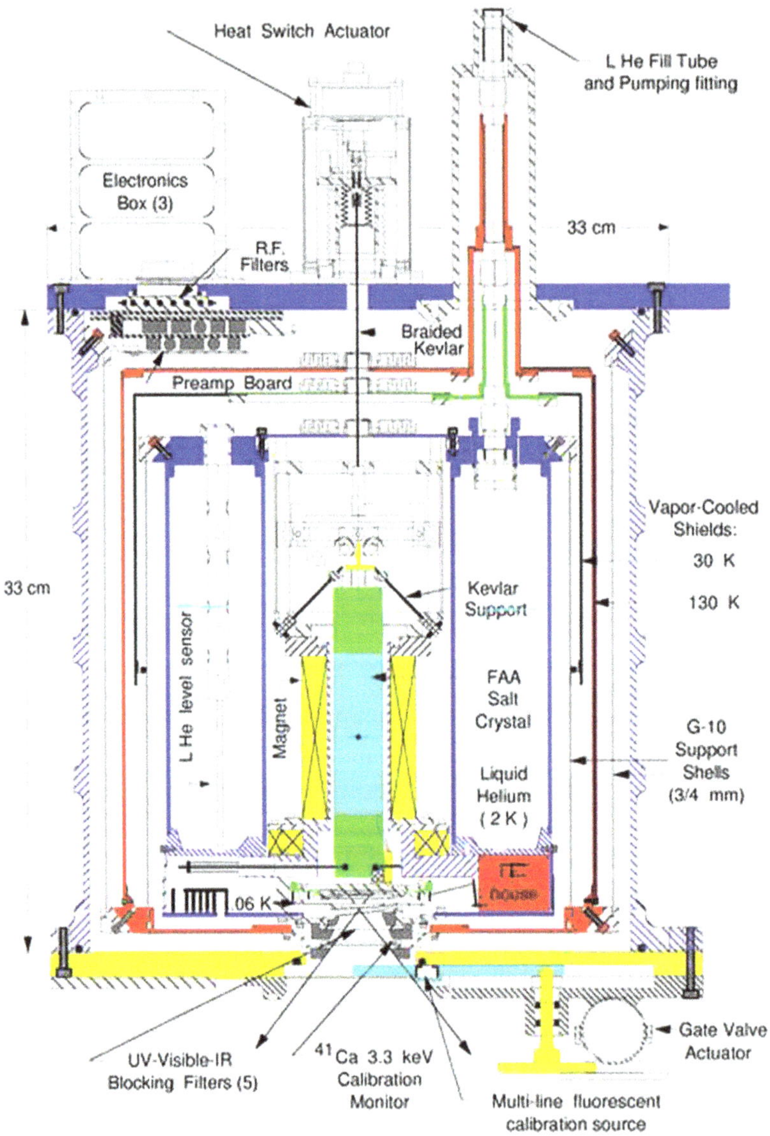

Figure 5.7. A diagram of an adiabatic demagnetization refrigerator developed by NASA. The paramagnetic material (salt crystal) is shown as the light blue material surrounded by the magnet. Courtesy of NASA (2001).

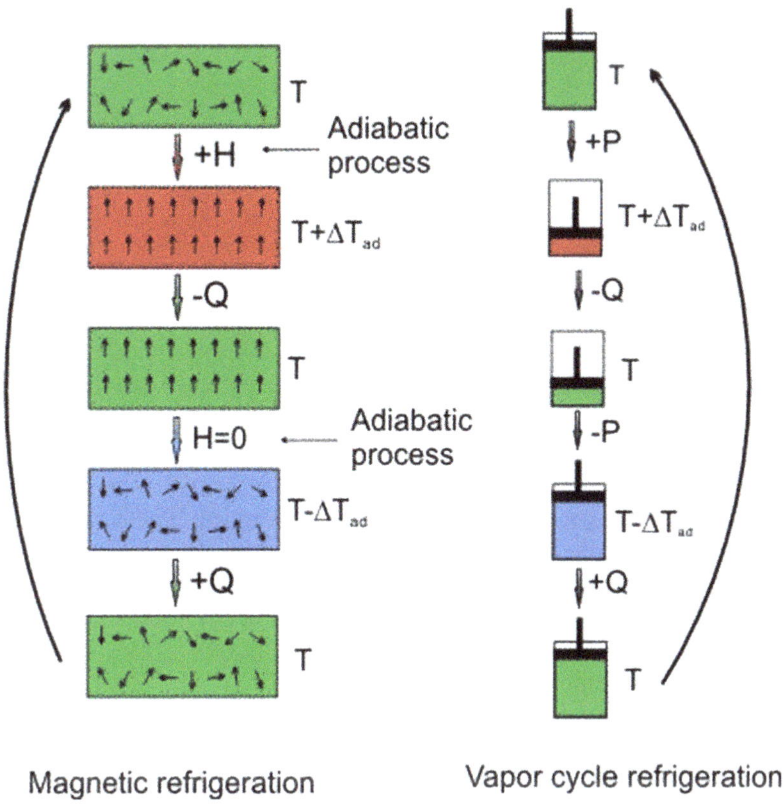

Figure 5.8. The basic operational principle of adiabatic demagnetization cooling and a comparison with conventional vapor-cycle refrigeration. From the top, the states of the system are: (1) the starting disordered magnetic state, (2) the state after adiabatic magnetization, (3) the state after isomagnetic enthalpic transfer, (4) the state after adiabatic demagnetization, and (5) the state after isomagnetic entropic transfer (which is the same as the original state). This Mozharivskyj (2004) image has been obtained by the author from the Wikimedia website, where it is stated to have been released into the public domain. It is included within this book on that basis.

H, is then applied to induce magnetization in the refrigerant. Since the magnetic state of the refrigerant is thereby changed from disordered to ordered, its entropy is decreased. Since the refrigerant is thermally isolated from its surroundings, the laws of thermodynamics require that its total energy remains unchanged. Hence, as its entropy decreases, its temperature increases to $T + \Delta T$.

Isomagnetic enthalpic transfer: The additional heat, Q, associated with the increase in temperature can be removed from the refrigerant by thermally coupling it to a heat sink by introducing a gas (i.e. helium) into the region around the refrigerant. During this process, the magnetic field is maintained so that the magnetic moments remain aligned.

Adiabatic demagnetization: The heat transfer gas is removed, thus thermally isolating the refrigerant once more, and the magnetic field is removed slowly, allowing the thermal energy to randomize the orientations of the magnetic moments. This process increases the entropy of the refrigerant and thereby decreases its temperature.

Isomagnetic entropic transfer: The refrigerant, now colder than it was at the beginning of the process, is thermally coupled to the experiment to be cooled, thereby removing heat and lowering its temperature. During this time, the magnetic field is held at zero; once the heat transfer is completed, the refrigerant is thermally decoupled from the experiment.

At the end of this process, heat has been removed from the experiment and the refrigerant has returned to its original magnetic state. The process can be repeated, removing additional heat from the experiment and cooling it further.

In principle, this process can be repeated indefinitely, which continues to cool the experiment. However, in practice, this does not work indefinitely. A major factor in the limitations of the adiabatic demagnetization refrigerator is the fact that paramagnetic materials are not ideal. At very low temperatures, the magnetic interactions between the magnetic moments become important. These interactions tend to align the magnetic moments, and the thermal energy is not sufficient to randomize their orientations, leading to a breakdown of the Curie's law behavior. Thus, during the adiabatic demagnetization process, the magnetic moments still show some residual preferential direction after the field is removed, thereby reducing the effectiveness of the cooling process.

The ultimate minimum temperature that can be achieved by this method depends on the ability of the refrigerant to avoid being subject to coupling between the magnetic moments of the atoms. $PrNi_5$ is one of the best-known refrigerants for this purpose and allows for cooling to temperatures of around 2 mK.

One way of cooling to lower temperatures than can be achieved with adiabatic demagnetization is to make use of adiabatic demagnetization associated with the very weak magnetic moments of the neutron and/or proton. If the atom has an electronic configuration in which all the electron spins cancel, then there is no atomic magnetic moment and the nuclear magnetic moment may be important. Such materials are referred to as diamagnetic materials. One example of a diamagnetic material is copper. A free copper atom has the electronic configuration $[Ar]3d^{10}4s^1$. In copper metal, the unpaired 4s electron exists in the conduction band and does not contribute to the localized atomic moment. This leads to the localized electronic configuration $[Ar]3d^{10}$, in which all the electrons are in filled shells. Natural copper consists of about 69% ^{63}Cu (with 29 protons and 34 neutrons) and 31% ^{65}Cu (with 29 protons and 36 neutrons). Therefore, all copper nuclei have one unpaired proton, leading to the formation of a net nuclear magnetic moment. This moment is, by analogy to equation (5.1), about three orders of magnitude smaller than a typical atomic magnetic moment because of the much greater proton mass compared with the electron mass. Thus, the magnetic coupling between the nuclear moments in diamagnets is much smaller than the coupling between atomic moments in paramagnets. This has both advantages and disadvantages. Obviously, the advantage is that nuclear adiabatic demagnetization can achieve much lower temperatures (i.e. μK) before the magnetic coupling affects the effectiveness of the method.

The disadvantage is that, because of the very weak moments associated with the nucleus, much larger applied magnetic fields are necessary in order to align the nuclear moments.

The approaches to cooling as described above impose a limit on the ultimate temperature that can be achieved. All these methods are similar in one, self-defeating, respect; they cool by transferring heat to a heat sink through a thermal link. The problem with this method is that heat can also flow into the sample from the outside through the thermal link. In order to reach the temperatures required for the experiments discussed in chapters 8 and 9, we need to take a different approach, and this requires looking at temperature in a different way.

5.4 Doppler cooling

It is clear that the atoms in a material are in motion as a result of thermal energy associated with them. We therefore commonly think of reducing this thermal motion by lowering the temperature. However, we could also think of lowering the temperature of a material by reducing the thermal motion of the atoms. This is the way we should look at temperature in order to understand how lasers can be used for cooling. In chapter 4, we looked at applications of laser light that depend on its different unique properties: monochromaticity, coherence, and intensity. It was obvious that the intensity associated with (at least some) lasers allows them to be used for heating. However, it is the monochromaticity of laser light that allows it to be used for cooling. In order to use a laser for cooling it is essential to be able to tune the wavelength (or energy) of the laser to the correct value. We begin with a brief discussion of tunable lasers.

In chapter 2, we saw that dye lasers are tunable over a wide range of wavelengths. In the early years of laser cooling experiments, dye lasers were the most appropriate tunable lasers that were available. In more recent years, other types of laser have been developed which can be tuned over a wide range of wavelengths. These include gas, solid-state, and semiconducting lasers.

Semiconducting lasers, or laser diodes, have a fairly broad distribution of wavelengths in their output, as described in chapter 3. It is also possible to adjust the output wavelength of these devices by changing their temperature, as their energy gap is a function of temperature. Their output can be tuned to a very narrow wavelength using a grating, along the lines of the scheme for tuning a dye laser shown in figure 2.12.

Probably the most useful, currently available laser that can be tuned over a wide range of wavelengths is the titanium–sapphire laser. In fact, the titanium–sapphire laser has the largest tunable range of any laser currently known (660–1180 nm). It is sometimes referred to as a Ti:sapphire or Ti:Al_2O_3 laser, as it consists of a solid-state lasing medium made of sapphire (Al_2O_3) with Ti impurities. This is analogous to the ruby laser, in which the lasing medium is sapphire with Cr impurities. The lasing transition in the Ti:Al_2O_3 laser occurs between the electronic states of the Ti^{3+} ion. The electronic configuration of the Ti^{3+} ion is [Ar]$3d^1$. The d-state is five-fold degenerate in m_l ($m_l = -2, -1, 0, +1, +2$). When the Ti^{3+} ion is contained in the host

Al_2O_3 matrix, the charge distribution from the neighboring atoms lifts this degeneracy. The local crystallographic environment of the Ti ion is an octahedron of six oxygen atoms, as illustrated in figure 5.9. Two of the m_l values (−2, +2) correspond to directions which represent bonds between the Ti and O ions along the z-axis, while the remaining three m_l values of (−1, 0, +1) do not correspond to bond directions between the Ti and O ions. The directions for $m_l = -2, +2$ represent a higher-energy configuration, and these form an excited doublet state. The other three values of $m_l = (-1, 0, +1)$ correspond to a lower-energy triplet state. This energy level splitting of the degenerate 3d level is shown in figure 5.10. The mean energy difference between the triplet ground state and the doublet excited state corresponds to a wavelength of about 500 nm, which falls in the green portion of the optical spectrum.

The energy levels of the Ti^{3+} ions are further complicated by the fact that the exact energies of the degenerate levels are a function of the position of the Ti ion inside the octahedral oxygen cage. A displacement of the Ti ion slightly lowers the energy of the excited doublet (the so-called Jahn–Teller effect) but not that of the ground-state triplet. Thermal vibrations of the Ti^{3+} ion in the octahedral cage produce changes in the Ti^{3+} energy levels, and these changes create phonons. The coupling of these phonons to the optical transitions smears the energy of the photons

Figure 5.9. The octahedral coordination of oxygen (green) atoms around a Ti ion (gray) in the Ti-substituted sapphire structure. magnetix / (Shutterstock.com).

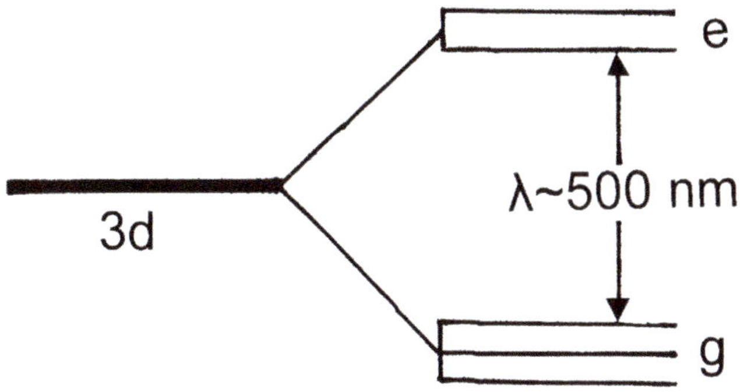

Figure 5.10. An energy level diagram of the Ti^{3+} ion showing the 3d ground state on the left and the split into an excited-state doublet and a ground-state triplet on the right.

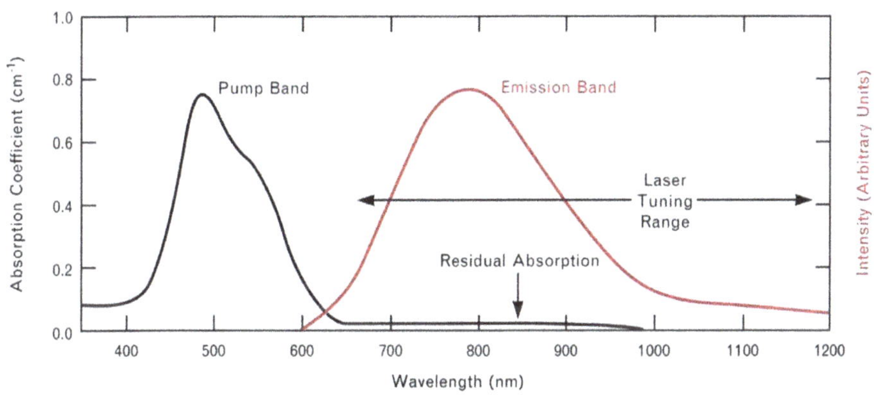

Figure 5.11. The optical spectra of the pumping radiation produced by an argon ion laser and the laser output of a Ti:Al_2O_3 laser. Reprinted from Wall and Sanchez (1990). Reprinted with permission of MIT Lincoln Laboratory, Lexington, MA, USA. Courtesy of MIT Lincoln Laboratory.

that are produced, giving rise to a very broad distribution of wavelengths. Because of this behavior, the Ti:Al_2O_3 laser is sometimes referred to as a vibronic laser.

In order to populate the excited state and create the necessary population inversion, the Ti:Al_2O_3 laser is optically pumped using higher-energy (i.e. shorter-wavelength) radiation that corresponds to the undistorted spacing of the energy levels, i.e. about 500 nm. The Ti:Al_2O_3 laser is most commonly optically pumped by an argon gas laser which has a mean output wavelength of just over 500 nm. This situation is illustrated in figure 5.11. The general design of a Ti:Al_2O_3 laser is illustrated in figure 5.12. The wavelength of the Ti:Al_2O_3 laser is tuned by adjusting the frequency of the master oscillator.

The ability of a laser to cool a collection of gas atoms relies on the interaction between the photons of the laser beam and the atoms and on our ability to control this interaction. This interaction may be described in terms of the properties of the photon. The photon obviously carries energy, but it also has momentum: $p = E/c = h/\lambda$, where E is the energy and λ is the wavelength. It is the conservation of

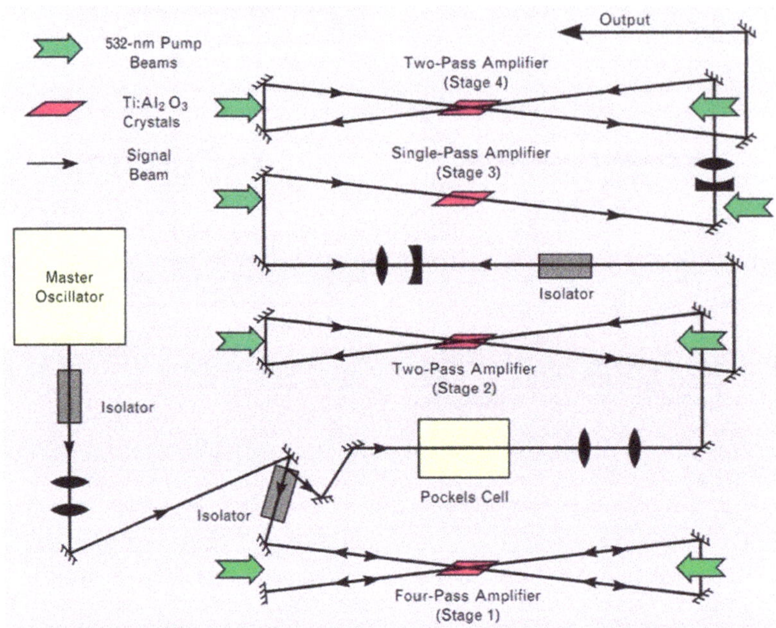

Figure 5.12. A diagram of the Ti:Al_2O_3 laser showing the pumping beam produced by an argon ion laser and the birefringent tuner. Reprinted from Wall and Sanchez (1990). Reprinted with permission of MIT Lincoln Laboratory, Lexington, MA, USA. Courtesy of MIT Lincoln Laboratory.

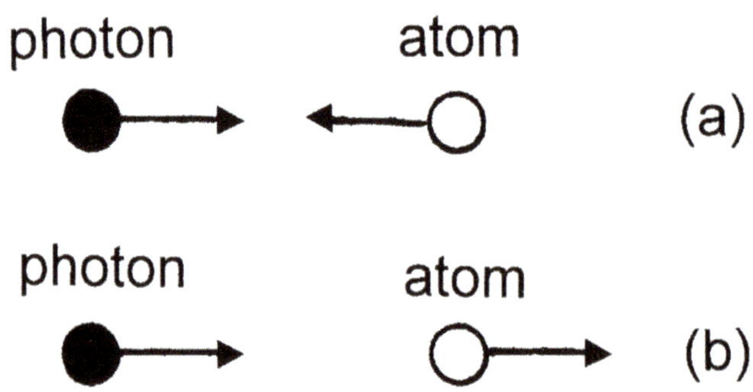

Figure 5.13. The interaction between a photon and an atom, in which (a) the momentum of the photon decreases the velocity of the atom and (b) the momentum of the photon increases the velocity of the atom.

momentum that describes the interaction. Let us look at a simple example. A laser beam is incident on a collection of gas molecules at a particular temperature. The gas molecules, by virtue of their temperature, are moving in random directions. We consider a simple one-dimensional view of this problem, in which the atoms are moving either parallel or antiparallel to the direction of the laser beam. First, consider a photon moving in the positive direction along the x-axis which encounters an atom moving in the negative direction as shown in figure 5.13(a). When the photon interacts with the atom, the momenta of the two particles are subtracted, leading to a reduction in the momentum (and hence the velocity) of the atom. In the opposite case, shown in figure 5.13(b), the momentum of the photon adds to the momentum of the atom, leading to an increase in the atom's velocity. A macroscopic manifestation of this basic physical behavior is referred to as radiation pressure. Although this simple example shows how we can affect the velocity of atoms using a laser, controlling the temperature of the gas is not so simple. This is because, in the gas, the atoms are moving at random and the effect of the laser beam is therefore to increase and decrease the velocity of atoms at random. To understand how we can avoid this problem, we need to look in more detail at the interaction between the photon and the atom.

Many experiments that make use of laser cooling techniques utilize alkali elements. These elements have a fairly simple electronic structure with filled shells and one additional s-state electron, e.g. Na is [Ne]3s, K is [Ar]4s, Rb is [Kr]5s, and Cs is [Xe]6s. This one extra s-electron has a fairly simple excited-state structure, in which the first excited state typically corresponds to the outer ns electron being excited into the np orbital. The s-electron can be excited from its ground state to the first excited state by irradiation with photons of just the right energy. This energy typically corresponds to photons in the visible region of the spectrum; thus, a tunable laser, as described above, is ideal for this purpose, as the photon energy can be adjusted to match the atom's excited-state energy. When the Heisenberg linewidth of the excited-state energy overlaps with the width of the wavelength envelope of the laser output, the laser radiation is readily absorbed and excites the atom into its excited state. This absorption slows the atom when it is approaching the photon at the time the absorption occurs, as the momenta partially cancel out. On the other hand, the absorption speeds the atom when it is moving away from the photon, as the momenta add. The way in which the atoms are preferentially slowed involves detuning the laser energy so that it is not centered on the energy of the atomic transition. Figure 5.14 illustrates this point.

Typically, the Heisenberg width of the atomic transition is several times the width of the energy envelope of the output produced by the laser. The laser energy is detuned (to a lower energy than the peak of the atomic transition energy) so that the center of the laser energy envelope sits about halfway down the low-energy side of the atomic transition energy distribution. The cooling effect described below is maximized when the peak in the laser energy occurs at the point of maximum slope in the curve representing the energy distribution of the atomic transition. The additional factor that we have not yet considered is the Doppler effect. The Doppler effect gives a shift in energy,

$$\Delta E = E_0\frac{v}{c},$$

when the source of photons is moving with a velocity v relative to the observer. When the atom is moving toward the source of photons (i.e. the laser), the photon energy is Doppler shifted to a higher value; that is, toward the peak in the energy distribution of the atomic transition. This increases the probability of photon

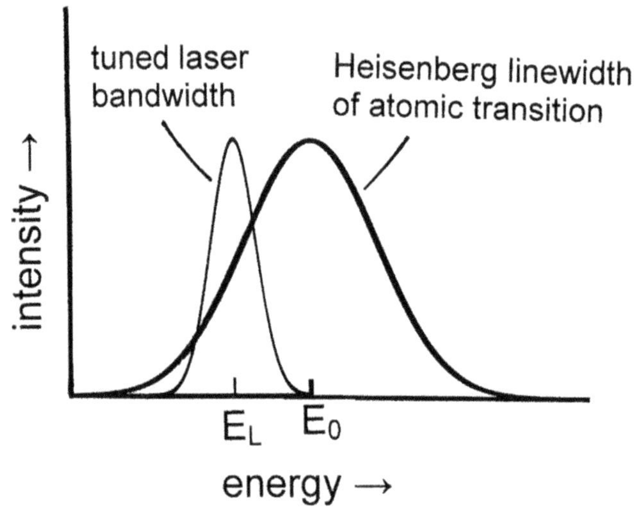

Figure 5.14. The detuning of laser energy to produce a suitable condition for Doppler cooling. E_L is the energy of the maximum in the laser output and E_0 is the center of the energy distribution associated with the atomic transition.

absorption and the probability that the atom's velocity will be affected and thereby decreased. When the atom is moving away from the source of photons, then the photon energy is Doppler shifted to a lower value, that is, away from the peak in the energy distribution of the atomic transition, and therefore the probability of absorption is decreased. Thus, although the velocity of the atom can still be increased or decreased through its interaction with the photon, it is more likely for this to occur when the atom is moving towards the laser. That means that more atoms have their velocity decreased than increased, leading to an overall decrease in the average atomic velocity and therefore a decrease in the temperature of the gas. In fact, the faster the atoms are moving, the greater the shift of the photon energy towards the peak in the transition energy and the greater the cooling effect. This technique is sometimes known as Doppler cooling.

Two final questions need to be considered at this point: what is the status of the atom after the cooling has occurred, and what happens to the kinetic energy that the atom has lost or the thermal energy that the gas has lost?

The atom is left in an unstable excited state after absorbing a photon. This unstable excited state spontaneously decays back to the ground state and, as a result, the atom emits a photon. The conservation of energy and momentum during this process requires that the atom recoil in the opposite direction to the direction of the emitted photon. Since the direction of the re-emitted photon is random (that is, it is unrelated to the direction of the initial photon that was absorbed), the recoil also takes place in a random direction. The isotropic re-emission means that the recoil sometimes speeds up the atom which was just slowed, but sometimes it further slows the atom. This effect averages to zero and, at least to a point, does not affect the overall temperature. We will see below that this is ultimately an important factor.

In general, the photon which was absorbed was at a lower energy than the peak in the energy distribution of the atomic transition. The re-emitted photon is, on average, at an energy that is equal to the peak energy in this distribution. Because of the random motion of the atoms, this re-emitted energy may sometimes be redshifted and may sometimes blueshifted by the Doppler effect relative to the peak energy, but on average, it is equal to the peak energy. So, the re-emitted photon is blueshifted (that is, at a higher energy) relative to the absorbed photon. The kinetic energy loss experienced by the atom in the process of slowing down from its interaction with the absorbed photon is exactly, on average, radiated away in the form of the excess energy associated with the re-emitted photon.

While it may seem that there would be no limit to the temperature that can be achieved by Doppler cooling, this is not the case. As suggested above, the re-emitted photons are important in this respect, as they represent the way in which the kinetic energy of the atoms is removed from the system. The ability to cool the gas by slowing the atoms is the result of shifting the laser energy toward the peak in the atomic energy distribution by the Doppler effect introduced by the thermal motion of the atoms. Thus, the relationship between the energy width of the atomic state, Γ, (as shown in figure 5.14) and the thermal energy of the atoms determines the low-temperature limit, T_D, to which the Doppler cooling method is effective. That is

$$k_B T_D = \frac{\Gamma}{2}$$

or

$$T_D = \frac{\Gamma}{2k_B}. \tag{5.3}$$

While Doppler cooling can lower the temperature of a system to a value well below those achieved by conventional cooling methods, the temperature given by equation (5.3) is typically around 100 μK, which is not sufficiently low to observe some of the phenomena that are described later in this book. Some common approaches to cooling below the Doppler limit are discussed in the next two sections. Forced evaporative cooling, which is a consequence of atomic trapping, is discussed in the next chapter.

5.5 Sisyphus cooling

The ground state of a free alkali atom, as noted above, has a magnetic dipole moment, μ, as a result of its unpaired s-electron. This atomic magnetic moment is antiparallel to the electron's spin (because of the electron's negative charge) and can be influenced by the application of an electric or magnetic field. The electromagnetic field associated with the laser beam lifts the degeneracy of the $s = 1/2$ ground state, yielding two hyperfine split states with $m_s = +1/2$ and $m_s = -1/2$. The magnitude of the split is related to the relationship between the spin direction and the polarization vector of the light. Two light beams of the same frequency which are in phase or π radians out of phase have a net linear polarization, while two beams that are $\pi/2$ or

$3\pi/2$ radians out of phase have net circular polarization. Two counterpropagating laser beams with opposite polarizations give rise to a spatially varying polarization that oscillates between left and right circular polarization. The spatial variations of the polarization state of the laser light give rise to a spatially varying split of the $m_s = +1/2$ and $-1/2$ state energies, as shown in figure 5.15.

Because the laser is detuned to the red for the purpose of Doppler cooling, the absorption of the photon, which slows the atom, always occurs from the higher sublevel of the ground state to the excited state of the atom. The re-emission of the photon which occurs during the decay back to the ground state is at a higher energy (as previously explained) and leaves the atom in the lower sublevel of the ground state. As an example, consider an atom traveling (say) from left to right in figure 5.15. If the atom is in the $m_s = -1/2$ state, then it preferentially absorbs a photon, inducing a transition into the excited state when the $-1/2$ sublevel has an energy near its maximum. When the excited state re-emits a photon, it has a greater energy than the absorbed photon, as described above, with the result that the atom ends up in the $m_s = +1/2$ sublevel of the ground state. These are the transitions represented by the left-hand set of arrows in the figure.

Some time later, the atom, which is now in the $m_s = +1/2$ sublevel, will be at the maximum energy; it can again absorb a photon and then decays back into the $m_s = -1/2$ state. This process is shown by the arrows on the right-hand side of figure 5.15. This process continues as the atom propagates to the right in the figure. Each time the atom climbs the energy curve of the ground-state sublevel, it converts kinetic energy into potential energy. During each cycle, this potential energy is radiated away by the increased energy of the re-emitted photon. Thus, the atom is continuously climbing up a potential hill, and the process is referred to as Sisyphus cooling after the character in Greek mythology who was condemned to repeatedly roll a huge boulder up a steep hill, only to see it roll back down to the bottom every time he got to the top.

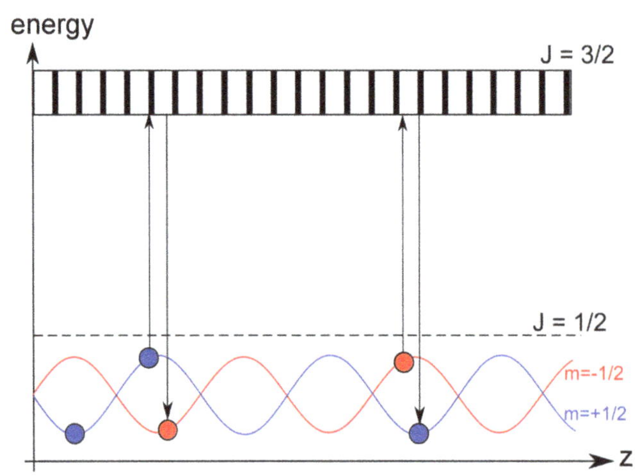

Figure 5.15. The principle of Sisyphus cooling as described in the text. Public domain by Stefan. This Stefan (2011) image has been obtained by the author from the Wikimedia website, where it is stated to have been released into the public domain. It is included within this book on that basis.

Sisyphus cooling continue to reduce the atom's thermal energy to the point where this energy becomes comparable to the energy difference between the m_s levels. As the split is a function of the laser light's intensity, decreasing the intensity of the laser lowers the ultimate temperature that can be obtained by this method. This is only possible up to a point, which brings us back to the question of the recoil energy of the atom. When the split becomes less than the thermal energy associated with an individual atomic recoil, Sisyphus cooling can no longer dissipate the energy that the atom acquires during the absorption/re-emission process. We can look at this quantitatively in the following way. The conservation of linear momentum equates the recoil momentum of the atom, p_R, to the momentum of the emitted photon

$$p_R = \frac{E_0}{c}.$$

The non-relativistic recoil energy is, therefore,

$$E_R = \frac{{p_R}^2}{2m} = \frac{{E_0}^2}{2mc^2},$$

where m is the atomic mass. Equating this energy to temperature leads to a theoretical lower limit for Sisyphus cooling of

$$T_S = \frac{{E_0}^2}{2k_B mc^2}.$$

For a typical alkali atom, this simple calculation gives a lower limit of about 2 μK. In practice, the lower limit that can be achieved experimentally is about an order of magnitude higher than this, but is still a factor of five or so lower than expected on the basis of simple Doppler cooling.

5.6 Other approaches to cooling below the Doppler limit

5.6.1 Anti-Stokes cooling

Anti-Stokes cooling can be utilized in systems with an appropriate energy level structure. Consider a system with three quantum levels. The ground state is designated $|\,0\rangle$ and two excited states are designated $|\,1\rangle$ and $|\,2\rangle$, as shown in figure 5.16. Figure 5.16(a) shows the situation in which a system in the ground state absorbs a photon and is excited up to state $|\,2\rangle$. This is followed by photon emission by spontaneous decay, either to state $|\,0\rangle$ (resonant scattering) or to state $|\,1\rangle$ (Stokes scattering). Figure 5.16(b) shows the situation in which a system in state $|\,1\rangle$ absorbs a photon and is excited up to state $|\,2\rangle$. This is followed by photon emission by spontaneous decay, either to state $|\,1\rangle$ (resonant scattering) or to state $|\,0\rangle$ (anti-Stokes scattering).

In the case of Stokes scattering, the emitted photon has less energy than the absorbed photon. In the case of anti-Stokes scattering, the energy of the emitted photon is greater than the energy of the absorbed photon; thus, the total energy (i.e. temperature) of the system is lowered. Anti-Stokes scattering competes with resonant scattering, and net cooling results when the equilibrium populations of

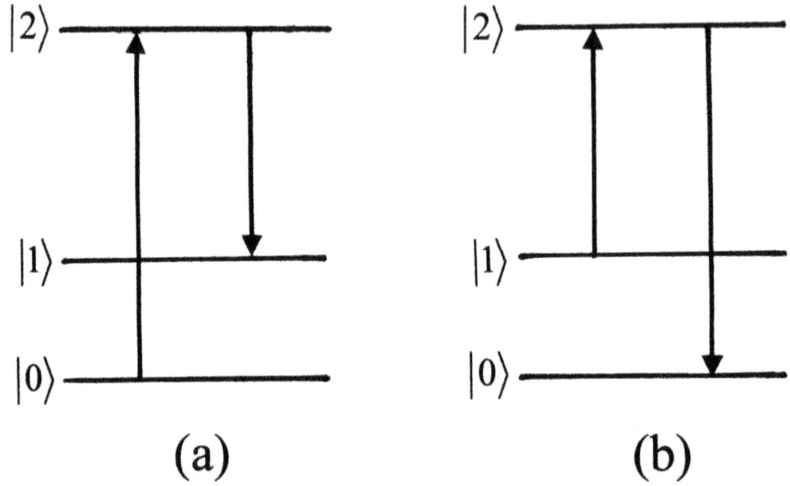

Figure 5.16. An energy level diagram of a three-level system showing (a) Stokes scattering and (b) anti-Stokes scattering, as described in the text.

states $|0\rangle$ and $|1\rangle$ are maintained mainly through non-radiative processes and the anti-Stokes transition rate is sufficiently large compared to the resonance transition rate. This latter factor can be expressed quantitatively in terms of the frequencies of the $|2\rangle \rightarrow |0\rangle$ and $|2\rangle \rightarrow |1\rangle$ transitions, ν_{20} and ν_{21}, respectively, and their transition rates, A_{20} and A_{21}, respectively. The cooling condition can therefore be written as

$$A_{20} > A_{21}\frac{\nu_{21}}{\nu_{20} - \nu_{21}},$$

and the cooling efficiency, ε, can be written as

$$\varepsilon = \frac{T_s}{T_a - T_s},$$

where T_s is the sample temperature and T_a is the surrounding ambient temperature.

From an experimental standpoint, a system which can be cooled by anti-Stokes scattering must have appropriate energy levels and there must be a suitable equilibrium population of the $|0\rangle$ and $|1\rangle$ states. Anti-Stokes cooling is then accomplished by an incident laser tuned to the appropriate frequency to initiate the $|1\rangle \rightarrow |2\rangle$ transition. The utilization of anti-Stokes scattering as a method of cooling was first proposed by Pringsheim (1929). It was first achieved experimentally by Djeu and Whitney (1981) for dilute gases of CO_2 and a mixture of CO_2–Xe. The first use of anti-Stokes cooling for the cooling of a solid sample was reported by Epstein *et al* (1995) for ytterbium-doped fluoride glass.

5.6.2 Dark-state cooling

Dark-state cooling, also referred to as gray molasses cooling, is similar to Sisyphus cooling in that it utilizes transitions between two hyperfine split ground states that are mediated by an excited state. It was first suggested by Weidemüller *et al* (1994) and was first observed experimentally by Boiron *et al* (1995). Unlike Sisyphus cooling, dark-state cooling utilizes a system in which only one of the hyperfine split ground states exhibits a spatially oscillating energy, as illustrated in figure 5.17. The higher-energy, oscillating ground-state sublevel is referred to as a 'bright' state and the lower-energy, non-oscillating sublevel is referred to as a 'dark' state. The combination of the bright state and the dark state gives rise to the terminology 'gray molasses.'

Dark-state cooling results from the laser excitation of the bright state into the excited state, followed by spontaneous emission to the dark state, as illustrated in figure 5.17. The excitation is caused by incident laser radiation that is tuned to the difference between the excited-state energy and the maximum energy of the bright state, so that atoms in the bright state undergo excitation when they have the maximum energy. After spontaneous emission from the excited state to the dark state, the dark-state atoms undergo a transition back to the bright state when the bright state is at a minimum in energy, as shown in the figure. Once in the bright state, the system follows the oscillatory energy curve back to the maximum, where the process can begin again. The repeated cycling of the system through these transitions carries away energy and cools the atoms. It is also interesting to note that

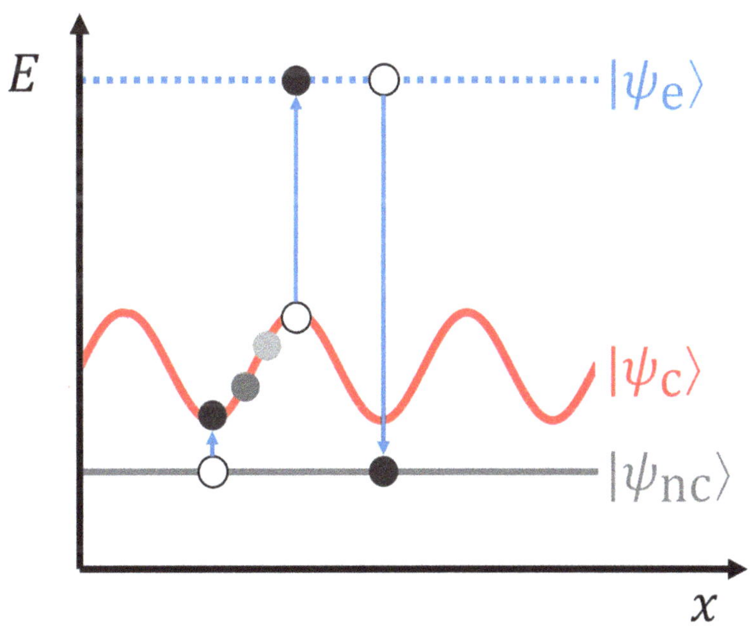

Figure 5.17. The energy transitions utilized in dark-state cooling as described in the text. The energy levels are shown as a function of position. For simplicity, spatial oscillations of the excited-state energy are not shown. This PercussiveMaintenance (2020) image has been obtained by the author from the Wikimedia website where it was made available under a CC BY-SA 4.0 licence. It is included within this book on that basis. It is attributed to PercussiveMaintenance (image cropped).

the transition from the dark state back to the minimum energy in the bright states occurs selectively for those atoms in the dark state which are near the top of their energy distribution. This ensures that the atoms which are left in the dark state are those which are at the lowest temperature.

From an experimental standpoint, dark-state cooling has been successfully implemented for atoms whose energy levels are not compatible with Sisyphus cooling in order to achieve temperatures below the Doppler limit. Some atomic species for which this approach has been utilized include isotopes of lithium, potassium, and cesium.

References and suggestions for further reading

Adwele 2012a Schematic diagram of a dilution refrigerator https://commons.wikimedia.org/wiki/File:Dilution_refrigerator01.jpg

Adwaele 2012b Cold part of dilution refrigerator https://en.wikipedia.org/wiki/File:Cold_part_of_dilution_refrigerator.jpg

Boiron D, Triché C, Meacher D, Verkerk P and Grynberg G 1995 Three-dimensional cooling of cesium atoms in four-beam gray optical molasses *Phys. Rev.* A **52** R3425(R)

Cao H 2021 Refrigeration below 1 Kelvin *J. Low Temp. Phys.* **204** 175–205

Djeu N and Whitney W T 1981 Laser cooling by spontaneous anti-Stokes scattering *Phys. Rev. Lett.* **46** 236–9

Epstein R, Buchwald M, Edwards B, Gosnell T R and Mungan C E 1995 Observation of laser-induced fluorescent cooling of a solid *Nature* **377** 500–3

Magnetix (n.d.) Influence of lone electron pairs on octahedral molecules - VSEPR model https://www.shutterstock.com/image-illustration/influence-lone-electron-pairs-on-octahedral-467129504

McClelland J J, Steele A V, Knuffman B, Twedt K A, Schwarzkopf A and Wilson T M 2016 Bright focused ion beam sources based on laser-cooled atoms *Appl. Phys. Rev.* **3** 011302

Metcalf H and van der Straten P 1999 *Laser Cooling and Trapping* (New York: Springer)

Mozharivskyj Y 2004 Analogy between magnetic refrigeration and vapor cycle or conventional refrigeration https://commons.wikimedia.org/wiki/File:MCE.gif

NASA 2001 XQC Payload https://phonon.gsfc.nasa.gov/rocket/payload/payload.html

PercussiveMaintenance 2020 An energy-level diagram of the gray-molasses cooling cycle, with position dependence https://commons.wikimedia.org/wiki/File:Gray_Molasses_Cooling_Cycle.png

Pérez A L, Neuman N I, Baggio R, Ramos C A, Dalosto S D, Rizzi A C and Brondino C D 2017 Exchange interaction between $S = 1/2$ centers bridged by multiple noncovalent interactions: contribution of the individual chemical pathways to the magnetic coupling *Polyhedron* **123** 404–10

Phillips W D and Cohen-Tannoudji C 1990 New mechanisms for laser cooling *Phys. Today* October 1990 33–40

Pringsheim P 1929 Zwei Bemerkungen über den Unterschied von Lumineszenz- und Temperaturstrahlung *Z. Phys.* **57** 739–46

Richardson R C and Smith E N (ed) ed. 1988 *Experimental Techniques in Condensed Matter Physics at Low Temperatures* (Reading, MA: Addison-Wesley)

Stefan 2011 Sisyphean cooling https://commons.wikimedia.org/wiki/File:Sisyphus.svg

Wall K F and Sanchez A 1990 Titanium sapphire lasers *Linc. Lab. J.* **3** 447–62

Weidemüller M, Esslinger T, Ol'shanii M A, Hemmerich A and Hänsch T W 1994 A novel scheme for efficient cooling below the photon recoil limit *Europhys. Lett.* **27** 109
White G K and Meeson P J 2002 *Experimental Techniques in Low Temperature Physics* 4th edn (Oxford: Oxford University Press)

Richard A Dunlap

Chapter 6

Laser trapping of atoms

6.1 Introduction

In the previous chapter, we discussed some common approaches to cooling to very low temperatures that utilize the interaction of atoms with laser radiation. In order to provide a useful environment in which to study the properties of systems at very low temperatures, it is necessary to not only cool the constituent atoms to a sufficiently low temperature but also to confine them spatially. This latter condition can be achieved by a combination of techniques, in which laser radiation plays an important role. This chapter summarizes the most significant of these techniques and also describes how laser light can be used to manipulate the position of atoms and how confinement can lead to further cooling of the system.

6.2 Optical traps

In section 5.4 it was shown how light from a laser could be used to reduce the velocity of atoms that are moving in a direction parallel to the beam. In the case of a single laser, the frequency of the laser is tuned so that atoms moving in the negative direction are preferentially slowed, while atoms moving in the positive direction are relatively unaffected. If two lasers are used, pointing at the collection of gas atoms from opposite directions, atoms moving in one direction are slowed by one laser and atoms moving in the opposite direction are slowed by the other laser. This behavior confines the atoms in the center of the space between the lasers. If three sets of lasers pointing in opposite directions are placed along three orthogonal axes, as shown in figure 6.1, then the atoms are spatially confined in three dimensions. Thus, the approach shown in figure 6.1 serves to both cool the atoms (as defined by the Doppler limit) and confine, or trap, the atoms; it is thus referred to as 'laser trapping.' Because the force slowing the atoms is proportional to their velocity, it

doi:10.1088/978-0-7503-5482-0ch6

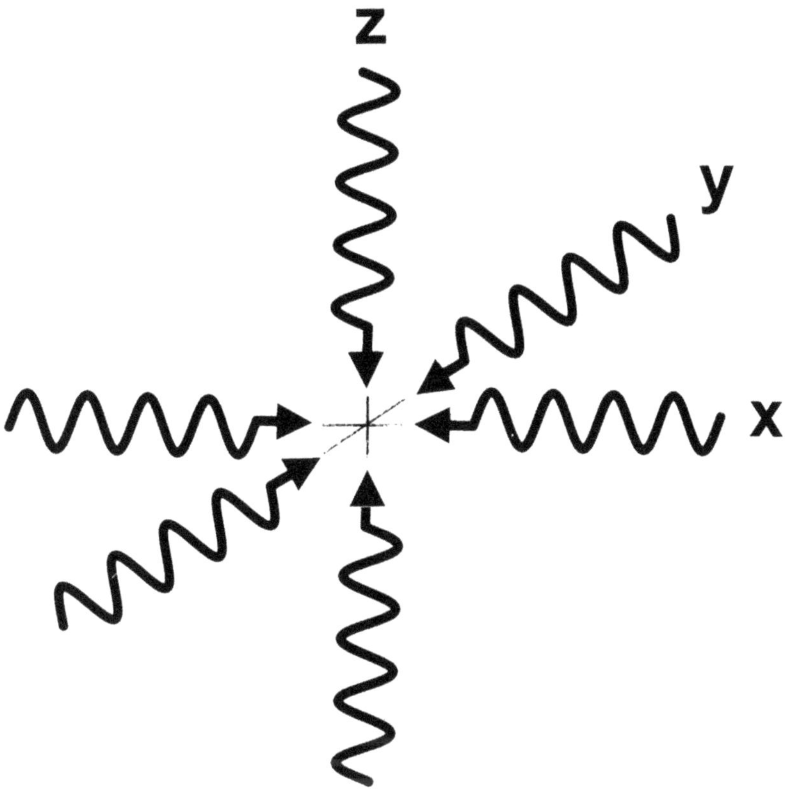

Figure 6.1. Confining atoms in three dimensions by laser trapping.

behaves rather like the viscous damping in a fluid. For this reason, this type of system is sometimes referred to as optical molasses.

6.3 Optical tweezers

Optical tweezers, which were originally referred to as a single-beam gradient force trap, were developed by Arthur Ashkin and co-workers (1970, 1986). They represent an experimental technique that utilizes a laser to trap and manipulate small particles. The technique can also provide a force that counteracts gravity, in which case the term 'optical levitation' is sometimes used.

A laser beam is focused to a very small diameter by means of lenses. The point of the beam that has the smallest diameter is referred to as the waist, as illustrated in figure 6.2. With proper optics, the limit to the radius of the waist, r, is given by

$$r = \frac{\lambda}{2} \tag{6.1}$$

where λ is the wavelength of the light. Focusing the light at the waist creates a very large electric field gradient, and this acts as to attract small dielectric particles to the point of the strongest electric field at the center of the beam. If the dielectric particle

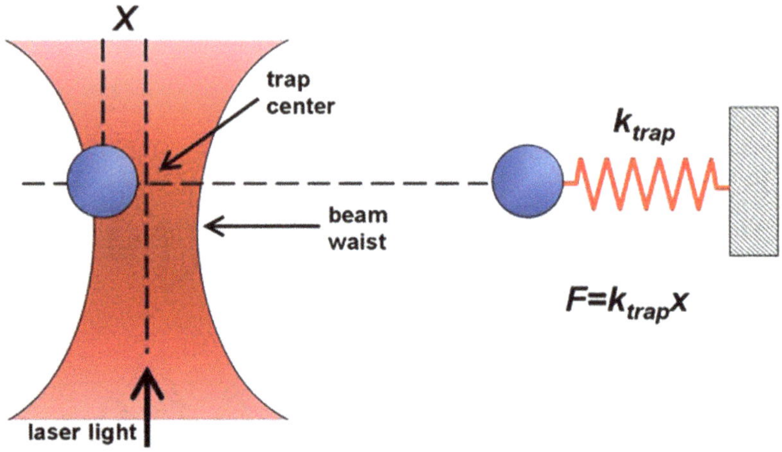

Figure 6.2. Description of optical tweezers. Public domain by RockyRaccoon. This [Optical trap as a spring] image has been obtained by the author from the Wikimedia website, where it is stated to have been released into the public domain. It is included within this book on that basis.

is displaced by a distance x from the center of the waist in a direction perpendicular to the propagation direction of the beam, then there is a restoring force that returns the particle to the point of the strongest field. To the first order, this restoring force, F, is described by Hooke's law in terms of a spring constant, k_{trap}, as

$$F = -k_{\text{trap}}x,$$

as shown in the figure.

The absorption or scattering of the laser light by the particle also exerts a force on the particle parallel to the direction of the beam propagation. This results from the conservation of the momentum carried by the optical photon and causes a displacement of the particle in the forward direction of the waist, as seen in figure 6.2. This displacement may be thought of as resulting from the radiation pressure caused by the laser light. The radiation pressure may be expressed in terms of the radiation force, F_{rad}, as

$$p = \frac{F_{\text{rad}}}{A}, \tag{6.2}$$

where the radiation force is given by the laser power, P, and the speed of light, c,

$$F_{\text{rad}} = \frac{P}{c}. \tag{6.3}$$

The area of the beam waist is given in terms of equation (6.1) as

$$A = \pi\left(\frac{\lambda}{2}\right)^2. \tag{6.4}$$

Combining equations (6.2), (6.3), and (6.4) gives

$$p = \frac{4P}{\pi c \lambda^2}.$$

The net equilibrium position of the particle in the beam direction is given in terms of the equilibrium between the radiation pressure and the restoring force that returns the particle to the point of the strongest electric field.

Optical tweezers may be used to trap dielectric particles and to manipulate their position by controlling the direction of the laser beam. A simple experimental arrangement for the use of optical tweezers to trap particles is illustrated in figure 6.3. This approach has been used to study a variety of small dielectric particles from the nanometer scale to the micron scale. An example of the optical observation of a silica nanoparticle trapped by this technique is shown in figure 6.4. A common application of the optical tweezer method has been the study of biological species, such as the interaction between DNA and protein molecules.

In 2001, Schlosser *et al* (2001) succeeded in trapping a single neutral atom using this approach. Figure 6.5 shows an example of single atom trapping. Here, two

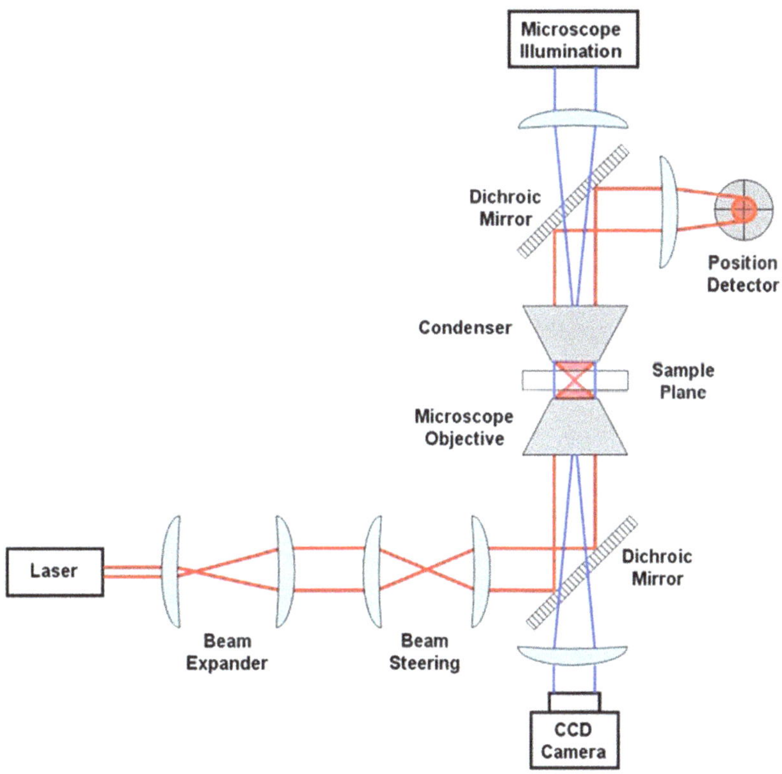

Figure 6.3. Diagram of a simple optical tweezer experiment. Public domain by RockyRaccoon. This [Generic optical tweezer diagram] has been obtained by the author from the Wikimedia website, where it is stated to have been released into the public domain. It is included within this book on that basis.

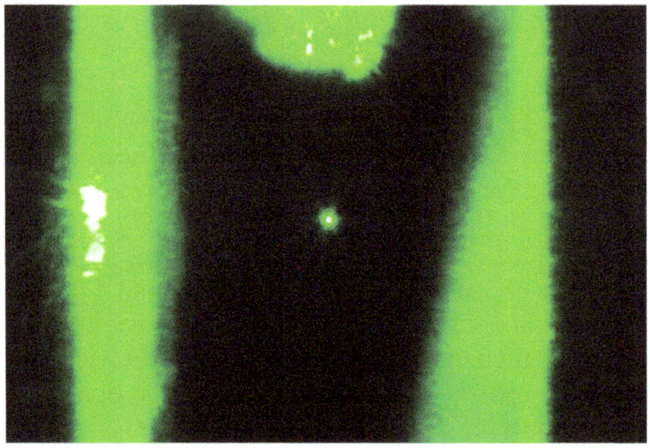

Figure 6.4. A silica nanoparticle suspended in an optical cavity. The green areas at the sides are the lenses used to focus the laser beam onto the nanoparticle. This [Levitated nanoparticle] image has been obtained by the author from the Wikimedia website where it was made available under a CC BY-SA 4.0 licence. It is included within this book on that basis. It is attributed to [StephTr].

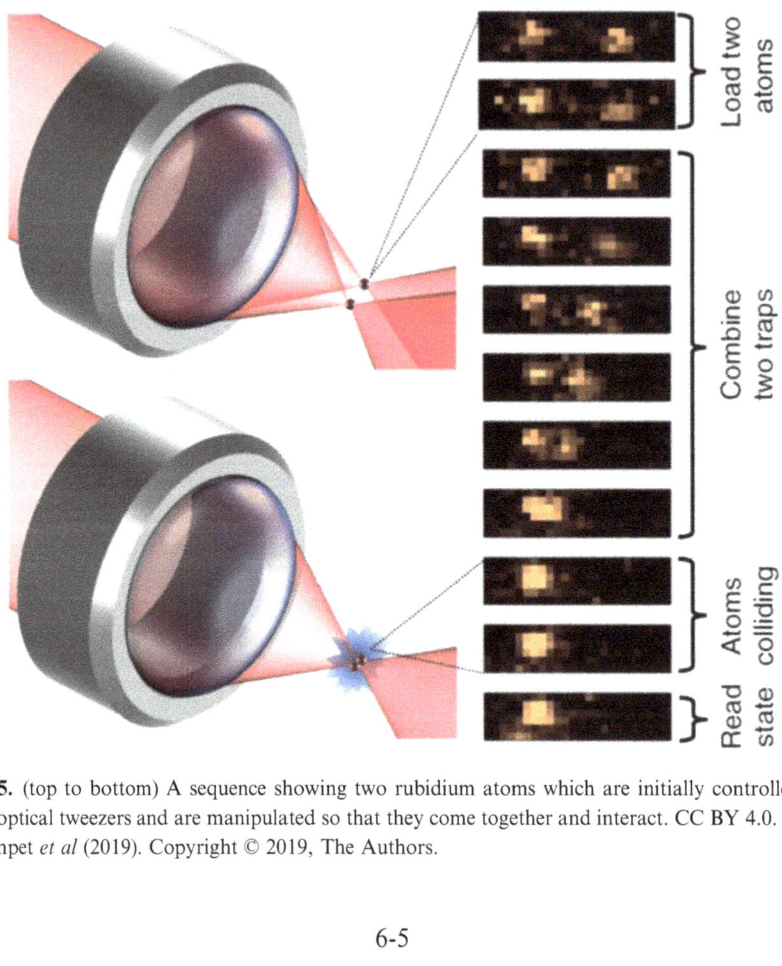

Figure 6.5. (top to bottom) A sequence showing two rubidium atoms which are initially controlled by two separate optical tweezers and are manipulated so that they come together and interact. CC BY 4.0. Reprinted from Sompet *et al* (2019). Copyright © 2019, The Authors.

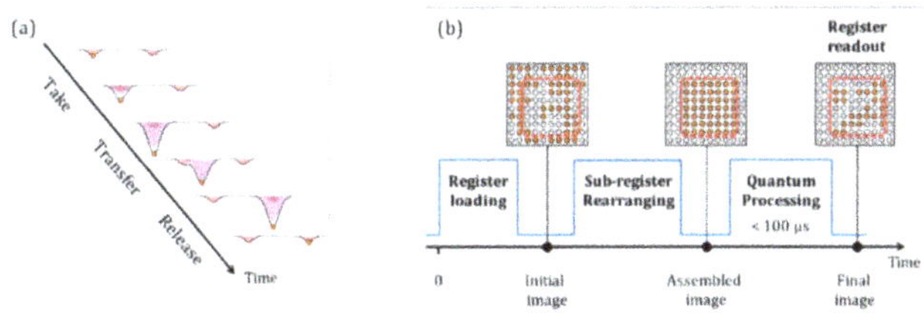

Figure 6.6. (a) The process of moving an individual atom from one trap to another using optical tweezers. The atom is first picked up from one site by the tweezer and then moved to another site where it is released. (b) Rearranging atoms in a register in order to store information. (left) The original arrangement of atoms in the register. (center) Rearranging atoms to form a defect-free register. (right) Arrangement of atoms to store information. Reprinted from Henriet *et al* (2020) CC BY 4.0.

rubidium atoms are trapped by two separate optical tweezers. The positions of the atoms are manipulated to allow the atoms to come into contact (Andersen 2022). These techniques allow few-atom systems to be assembled from individual atoms. The quantum mechanics of few-body systems can be studied without the interference of the surrounding environment. The ability to manipulate individual atoms with optical tweezers is also an important component of possible quantum information systems (see, for example, Henriet *et al* 2020, Saffman *et al* 2010). Manipulating individual atoms using optical tweezers allows for the storage of information in an array of traps. This can form the basis of a quantum computer, and the fundamentals of this technology are illustrated in figure 6.6. Further details of the formation of traps appropriate for such information storage have been presented by Weitenberg *et al* (2011) and are discussed in chapter 9.

6.4 Magnetic traps

The discussion of laser cooling in the previous chapter showed how the velocity of the atoms, and hence their temperature, was lowered. This approach confined the atoms to a point near the origin in velocity space, i.e. $v_x = v_y = v_z = 0$. This confinement in velocity space does not necessarily confine the atoms to a specific location in real space. This is particularly important in the z-direction, as the atoms are subject to the gravitational force. The atoms may be spatially confined by the use of magnetic fields. Magnetic confinement results from the force, $\boldsymbol{F}$, acting on a magnetic moment, $\boldsymbol{\mu}$, in a magnetic field gradient, given by

$$\boldsymbol{F} = \nabla(\boldsymbol{\mu} \cdot \boldsymbol{B}),$$

where $\boldsymbol{B}$ is the magnetic flux density. This magnetic confinement in space, combined with the velocity confinement achieved through the use of Doppler cooling, is referred to as a magneto-optical trap. This section describes the physical principles of magnetic traps, and the following section shows how magnetic traps are used in conjunction with Doppler cooling.

As discussed above, the magnetic moment associated with a free alkali atom couples to magnetic fields and can therefore be influenced not only by the electromagnetic field associated with a laser beam but also by an externally applied magnetic field. The magnetic moment can align parallel or antiparallel to the magnetic field, thereby respectively decreasing or increasing the atom's energy. The energy difference between the two spin alignments is

$$\Delta E = 2\mu B, \tag{6.5}$$

where the lowest energy is achieved by a parallel alignment of the magnetic moment and the applied field. The use of anti-Helmholtz coils, as discussed below, is ideal for creating a region of very low magnetic field where the magnetic split is minimized and atoms can be trapped.

Helmholtz coils are in common use as a means of filling a fairly large volume of space with a fairly uniform magnetic field. The coils consist of two thin solenoids, each with a radius *R*, which are separated by a distance *R*, as illustrated in figure 6.7. In the normal Helmholtz configuration, the currents in the two solenoids travel in the same direction and the two coils produce magnetic fields with a high degree of uniformity which point in the same direction and add together at the centers of the solenoids. These are the locations where the field is the largest and the most uniform.

Anti-Helmholtz coils use the same geometry as Helmholtz coils, but the direction of the current in one of the solenoids is reversed. This produces a region of zero

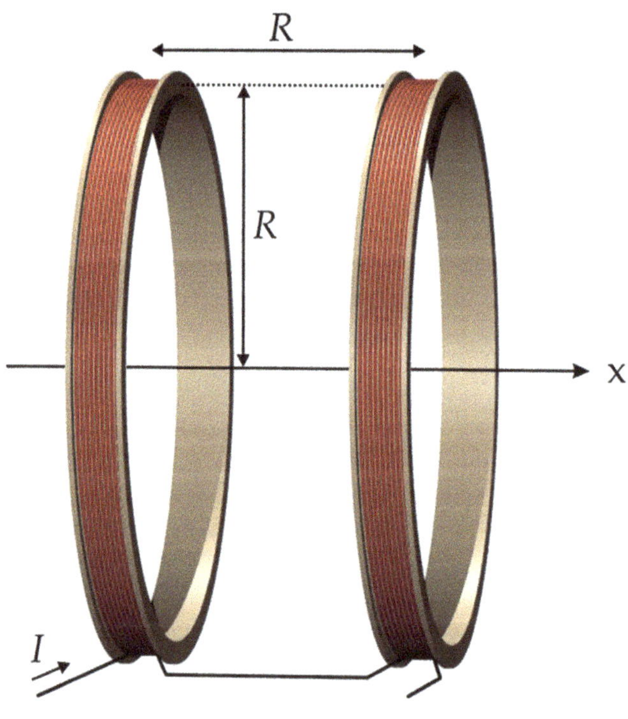

Figure 6.7. Helmholtz coil geometry. This [Helmholtz coils] image has been obtained by the author from the Wikimedia website where it was made available under a CC BY-SA 2.0 licence. It is included within this book on that basis. It is attributed to Ansgar Hellwig.

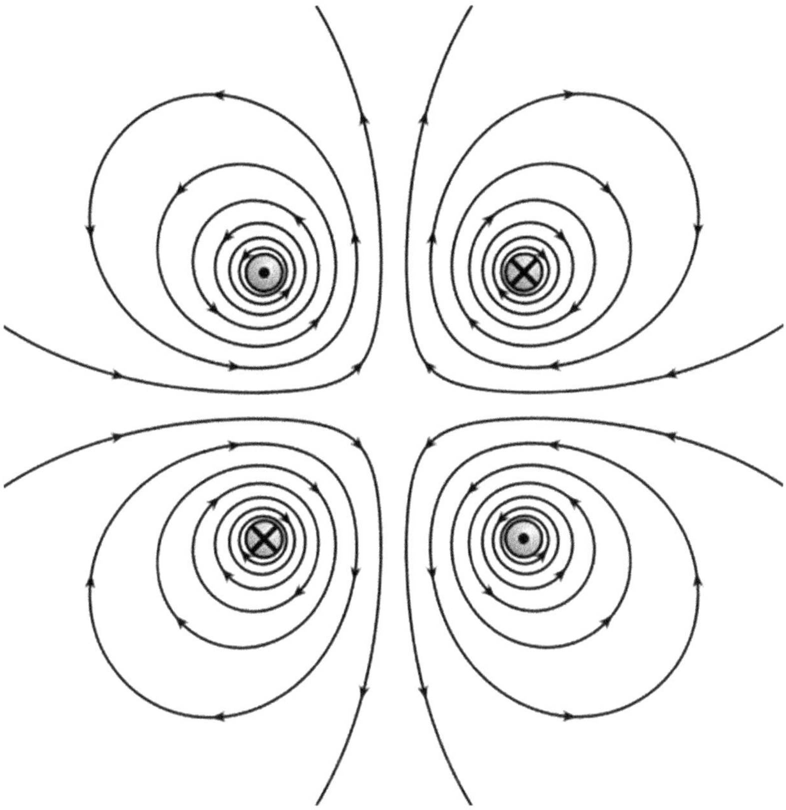

Figure 6.8. The field lines of a magnetic quadrupole trap. This [Magnetic field of four infinite wires perpendicular to the image plane] image has been obtained by the author from the Wikimedia website where it was made available under a CC BY-SA 3.0 licence. It is included within this book on that basis. It is attributed to Geek3.

magnetic field at the center, with a magnetic field of increasing magnitude in all directions away from the center. The magnetic field produced by a single solenoid (or a bar magnet) is a dipole field, like the magnetic dipole moment produced by an electron's orbital angular momentum. The field produced by an anti-Helmholtz coil is a quadrupole field, and the zero-field region at the center forms a magnetic quadrupole trap, as shown in figure 6.8. Commonly, for atomic trapping experiments, anti-Helmholtz coils are oriented so that their axes are aligned in the z-direction (i.e. the gravitational field direction).

6.5 Magneto-optical traps and forced evaporative cooling

A magneto-optical trap, as shown in figure 6.9(a), may be created by combining a laser trap (as shown in figure 6.1) that confines the atoms to near-zero velocity in three dimensions with a magnetic trap (created by anti-Helmholtz coils) that confines the atoms spatially.

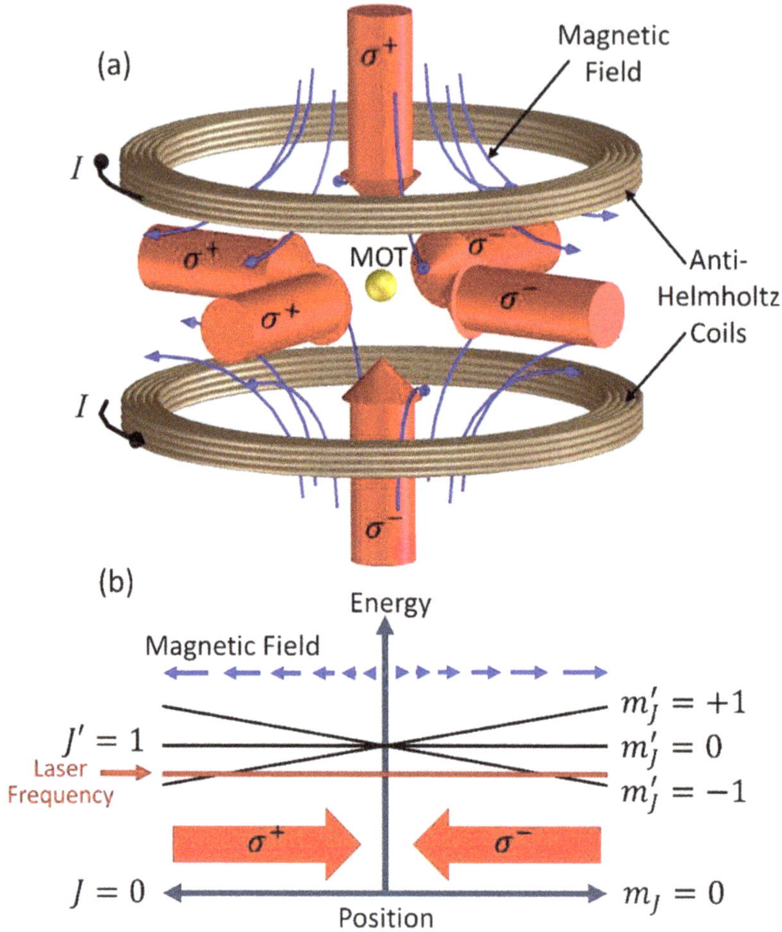

Figure 6.9. The magneto-optical trapping of atoms and spin-flip forced evaporative cooling. (a) A magneto-optical trap (MOT) formed by the combination of three mutually orthogonal laser beams and a magnetic field gradient produced by anti-Helmholtz coils. (b) An energy level diagram of magnetically split energy levels around the magneto-optical trap. Reprinted from McClelland *et al* (2016) with the permission of AIP Publishing.

The combination of the cooling methods described in the previous chapter can lower the temperature to a few μK. This is still too high for a number of experiments, such as Bose–Einstein condensation, as described later in this book. Additional cooling, which is the final step for many ultralow-temperature experiments, uses a very simple and well-known technique, that of evaporative cooling. For a classical system, the velocity distribution is given by the Maxwell–Boltzmann distribution, as illustrated in figure 1.10. It is the particles with the greatest velocity in the high-velocity tail of the distribution that are lost to evaporation. These high-energy particles carry away a disproportionately large fraction of the energy, because the energy per particle is proportional to the square of the velocity. Thus, the evaporation of a relatively small fraction of the particles in a system can give rise to a substantial reduction in temperature. It is essential in this cooling process that once the high-velocity particles have been lost due to evaporation, the remaining particles re-equilibrate by distributing their energies to create a new proper Maxwell–Boltzmann distribution for the new temperature. Below, we look at how this approach can be used effectively at very low temperatures.

Since the laser cooling technique limits the ultimate temperature that can be achieved, it is necessary to temporarily turn off the lasers to allow the system to cool to a lower temperature. The gas atoms are now just confined within the magnetic trap. In the region of very low magnetic field created by the anti-Helmholtz coils, the magnetic split given by equation (6.5) is minimized and increases spatially in all directions from the center of the trap, as shown in figure 6.9(b). The lower sublevel in the figure corresponds to a parallel alignment of the atom's magnetic moment and the applied magnetic field, while the upper sublevel corresponds to an antiparallel alignment of the moment and the applied field. It is clear from the figure that those atoms with moments that align antiparallel to the field are in a stable energy minimum and are trapped, while those with a parallel alignment are not. Transitions between the lower and upper states can be induced by the application of a radio-frequency (rf) magnetic field with an appropriate angular frequency, ω, such that

$$\hbar\omega = 2\mu B.$$

Controlling the duration of the rf pulses allows the atoms to be configured in either desired state. The atoms with antiparallel moments and the lowest velocities sit near the minimum in the upper sublevel in the middle of the magnetic trap. Those atoms with antiparallel moments and higher velocities climb up the potential and reside in the outer regions of the magnetic trap. By applying a tuned rf pulse of the proper frequency and duration, these higher-energy atoms near the edges of the trap can undergo a spin flip and fall down into the lower parallel moment sublevel. Once there, it is energetically favorable for them to fall off the edge of the energy curve and evaporate from the trap, carrying energy away with them. Since it is the higher-velocity atoms from the antiparallel level that are removed, the atoms that remain in the trap re-equilibrate to a lower temperature. This process can be repeated, lowering the temperature at each step at the expense of losing a small fraction of the atoms in the trap. This final step of cooling by forced evaporation can lower the temperature to the nK range.

References and suggestions for further reading

Andersen M F 2022 Optical tweezers for a bottom-up assembly of few-atom systems *Adv. Phys.: X* **7** 2064231

Ashkin A 1970 Acceleration and Trapping of Particles by Radiation Pressure *Phys. Rev. Lett.* **24** 156–9

Ashkin A, Dziedzic J M, Bjorkholm J E and Chu S 1986 Observation of a single-beam gradient force optical trap for dielectric particles *Opt. Lett.* **11** 288–90

Cao H 2021 Refrigeration below 1 Kelvin *J. Low Temp. Phys.* **204** 175–205

Geek3 2010 Magnetic field of four infinite wires perpendicular to the image plane, conducting current into and out of the image plane. https://commons.wikimedia.org/wiki/File:VFPt_four_wires_antiparallel.svg

Hellwig A 2005 Helmholtz coils https://commons.wikimedia.org/wiki/File:Helmholtz_coils.png

Henriet L, Beguin L, Signoles A, Lahaye T, Browaeys A, Reymond G O and Jurczak C 2020 Quantum computing with neutral atoms *Quantum* **4** 327–60

McClelland J J, Steele A V, Knuffman B, Twedt K A, Schwarzkopf A and Wilson T M 2016 Bright focused ion beam sources based on laser-cooled atoms *Appl. Phys. Rev.* **3** 011302

Metcalf H J and Straten P 1999 *Laser Cooling and TrappingGraduate Texts in Contemporary Physics* (New York: Springer)

Pérez A L, Neuman N I, Baggio R, Ramos C A, Dalosto S D, Rizzi A C and Brondino C D 2017 Exchange interaction between *S*= 1/2 centers bridged by multiple noncovalent interactions: contribution of the individual chemical pathways to the magnetic coupling *Polyhedron* **123** 404–10

Phillips W D and Cohen-Tannoudji C 1990 New mechanisms for laser cooling *Phys. Today* **43** 33–40

Richardson R C and Smith E N (ed) 1988 *Experimental Techniques in Condensed Matter Physics at Low Temperatures* (Reading, MA: Addison-Wesley)

RockyRacoon 2007a Optical Trap As a Spring https://en.wikipedia.org/wiki/File:Optical_Trap_As_a_Spring.jpg

RockyRacoon 2007b Generic Optical Tweezer Diagram https://commons.wikimedia.org/wiki/File:Generic_Optical_Tweezer_Diagram.jpg

Saffman M, Walker T G and Mølmer K 2010 Quantum information with Rydberg atoms *Rev. Mod. Phys.* **82** 2313–63

Schlosser N, Reymond G, Protsenko I and Grangier P 2001 Sub-poissonian loading of single atoms in a microscopic dipole trap *Nature* **411** 1024–7

Sompet P, Szigeti S S, Schwartz E, Bradley A S and Andersen M F 2019 Thermally robust spin correlations between two ^{85}Rb atoms in an optical microtrap *Nat. Commun.* **10** 1889

StephTr 2023 A silica nanoparticle trapped in an optical tweezer, illuminated by a green laser beam https://commons.wikimedia.org/wiki/File:Levitated_nanoparticle.jpg

Wall K F and Sanchez A 1990 Titanium sapphire lasers *Linc. Lab.* J **3** 447–62

Weitenberg C, Kuhr S, Mølmer K and Sherson J F 2011 Quantum computation architecture using optical tweezers *Phys. Rev.* A **84** 032322

White G K and Meeson P J 2002 *Experimental Techniques in Low Temperature Physics* 4th edn (Oxford: Oxford University Press)

Richard A Dunlap

Chapter 7

Fermions and bosons

7.1 Introduction

In the first part of this book, we have seen some of the properties of particles in interacting systems. These properties include the classical behavior of particles in a gas, as described by the Maxwell–Boltzmann distribution in figure 1.10, and the quantum mechanical behavior of electrons in a solid. We have also seen that the behavior of systems of particles is greatly influenced by temperature. While the majority of materials used for commercial applications are used at around room temperature, many systems have low-temperature behaviors that are of considerable scientific interest. Bose–Einstein condensation is one of the phenomena which occurs in some systems at very low temperatures; it is an important example of the application of laser cooling and trapping techniques described in the last two chapters. We begin this chapter with an overview of the two different types of particle that exhibit quantum mechanical behavior, namely fermions and bosons, and their properties as a function of temperature. The next chapter describes the behavior of bosons at the very low temperatures at which Bose–Einstein condensates are formed.

7.2 Fermions, bosons, and the Pauli principle

Electrons must obey the Pauli exclusion principle, and this requirement determines their ability to occupy various quantum levels in a system. More precisely, the Pauli principle requires that each electron in an interacting system must have a unique set of good quantum numbers. The electrons in an atom occupy quantum mechanical energy levels designated by a principal quantum number n. In a multielectron atom, the electron–electron interactions split the levels defined by n into sublevels designated by the angular momentum quantum number l, where $l < n$. A level described by the orbital angular momentum number l is degenerate in the

doi:10.1088/978-0-7503-5482-0ch7

z-component of the orbital angular momentum with a degeneracy of $2l+1$. That is, there are $2l+1$ possible orientations of the orbital angular momentum vector, as shown by the example for $l = 2$ in figure 7.1.

An electron distinguished by the set of quantum numbers, n, l, and m_l, is also described by a spin quantum number s, where the intrinsic spin of the electron is $s = 1/2$. The spin has two possible orientations defined by its z-component m_s, where $m_s = -1/2$ or $m_s = +1/2$. This gives a total degeneracy of an l level of $2(2l + 1)$. Thus, a unique set of good quantum numbers n, l, m_l, and m_s, defines each electron in an atom and determines how electrons occupy the available quantum energy levels.

We also saw in chapter 3 how electrons can be thermally excited into unoccupied higher-energy levels. This behavior can be quantitatively explained by the Fermi–Dirac function:

$$f(E) = \frac{2}{\exp\left[\dfrac{E-\mu}{k_B T}\right] + 1},$$

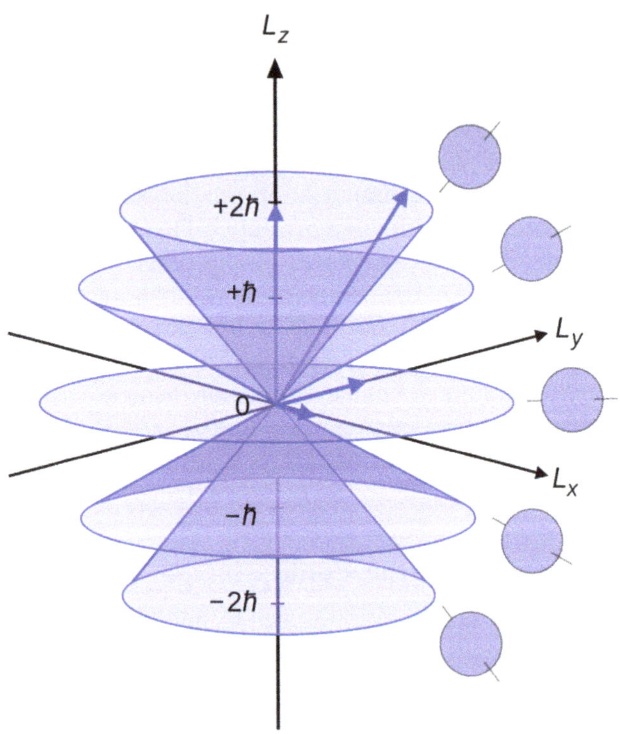

Figure 7.1. Cones defined by the possible values of the z-component, m_l, of the orbital angular momentum quantum number, l, for $l = 2$. This [Vector model of orbital angular momentum] has been obtained by the author from the Wikimedia website, where it is stated to have been released into the public domain. It is included within this book on that basis.

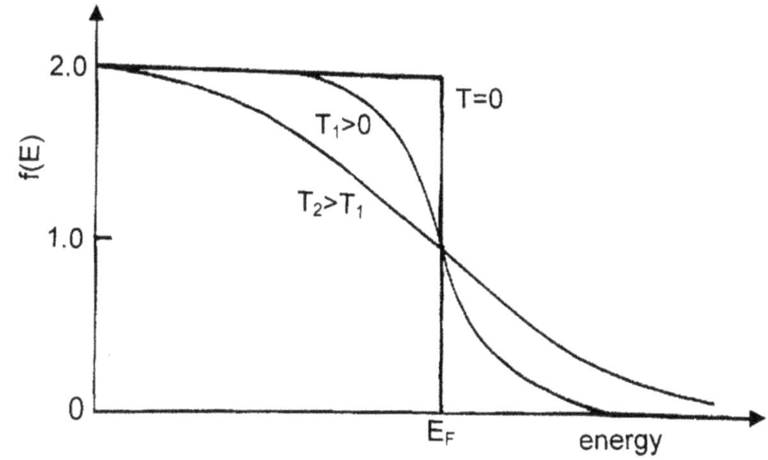

Figure 7.2. Fermi–Dirac distributions at different temperatures. The Fermi energy, E_F, is the energy separating occupied states and unoccupied states at $T = 0$.

as illustrated in figure 7.2. Here, μ is the chemical potential (which reduces to the Fermi energy shown in the figure at zero temperature) and k_B is the Boltzmann constant. The factor of two in the numerator arises from the spin degeneracy.

The behavior of electrons follows the description given above because electrons are fermions. A fermion is any particle or system of particles that has a half-integer spin, i.e. 1/2 in the case of the electron. Other fermions include the proton, the neutron and atoms that have a net half-integer spin. For a particle with a spin of s, the spin angular momentum is given in SI units by $\hbar s$, where $\hbar$ is the Planck constant. It is interesting to note that s is dimensionless, so the Planck constant has units of angular momentum, J s.

We do not have a real fundamental understanding of why particles with half-integer spin must obey the Pauli principle and follow Fermi–Dirac statistics. However, it is known that particles and systems of particles with integer spin do not behave in this way. These particles, known as bosons, after the Indian physicist Satyendra Nath Bose, are exempt from the Pauli principle and are described by Bose–Einstein statistics. In 1924, Bose suggested this model to describe the behavior of photons and it was generalized for all integer spin particles by Einstein the following year. Therefore, Bose–Einstein statistics describe the behavior of photons (spin 1) as well as all other particles or systems of particles with integer spin. The Bose–Einstein distribution has the temperature dependence

$$f(E) = \frac{g}{\exp\left[\dfrac{E-\mu}{k_B T}\right] - 1}, \tag{7.1}$$

where g is the state degeneracy. Figure 7.3 shows the Bose–Einstein distribution at a finite temperature. A comparison of figures 7.2 and 7.3 shows a fundamental difference between Fermi–Dirac statistics and Bose–Einstein statistics that arises

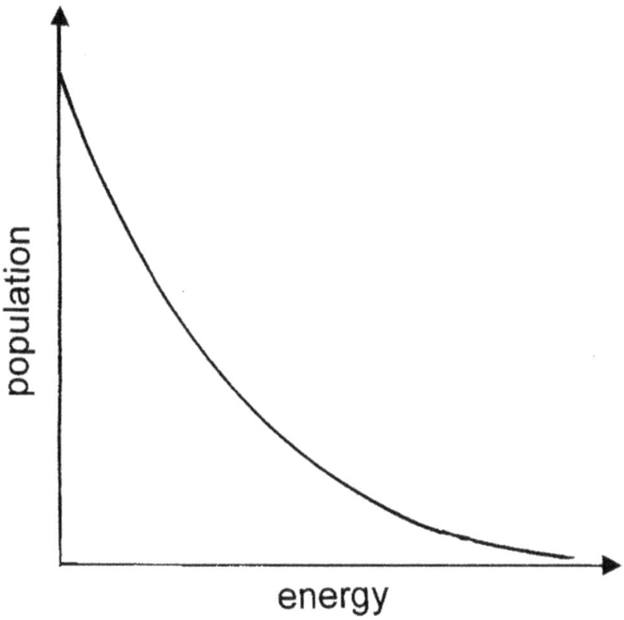

Figure 7.3. The Bose–Einstein distribution at a finite temperature, as given by equation (7.1).

because of the difference in the way that the Pauli principle relates to fermions and bosons.

As the temperature of a system of bosons approaches absolute zero, the form of the Bose–Einstein distribution shows that all particles occupy the same ground state at $E = 0$. This is quite distinct from the ground state for a system of fermions, for which figure 7.3 shows that the distribution is a step function which goes to zero at the Fermi energy. It is this behavior of bosons in their ground state that gives rise to the very unusual phenomenon of Bose–Einstein condensation. A fundamental concept, which forms the basis of our understanding of why the ground state of a system of bosons behaves as it does, is the inability to distinguish between particles in a quantum system, as discussed in the next section.

7.3 Distinguishable and indistinguishable particles and quantum states

Consider two macroscopic objects that are readily identifiable in a system consisting of two states of equal energy. Figure 7.4 shows a simple example of this situation, in which a white ball and a black ball can exist in one of two boxes. It is clear that there are four possible configurations for the system, as illustrated in the figure, each with equal probability. If we replace the objects in the example with identical objects, e.g. two visually identical balls, there are still four possible states, each with equal probability. This is because the objects may be identical, but they are not indistinguishable. If we observe the system closely, we can keep track of which

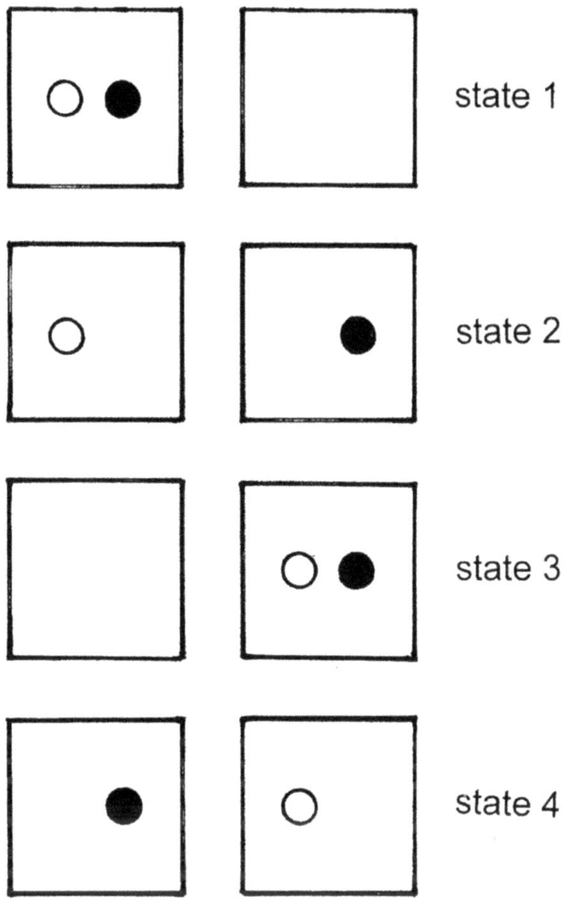

Figure 7.4. The state configurations of two distinguishable objects in two states with equal energy.

ball is in which box and we can observe changes of state by observing which ball moves where.

If we now consider a macroscopic system in which there are two states and N distinguishable objects, then we can calculate that the number of distinct configurations of the system is 2^N, since each of the objects has an equal probability of being in either of the two states and the probability of each configuration is the same. Since there are more configurations which have approximately equal numbers of objects in each box, it is most likely that the objects will be more or less equally divided between the two boxes.

If we consider the same problem for identical particles that must obey quantum mechanics, then we find that the results are quite different because such objects are indistinguishable. Indistinguishability is a feature that results from the Heisenberg uncertainty principle. It is not possible to follow the trajectory of a particle, so if a particle changes state, it is not possible to know which particle it is. In the case of two particles and two states, there are only three (rather than four) possible states; one particle in each state, two particles in one state or two particles in the other state. It is not important which particle is where, only how many particles are in each state. If we consider a system of N particles and two states, then the number of configurations is merely $N + 1$, each with equal probability, and the only relevant factor is how many particles are in each state and not which particles are where. Thus, the number of particles in one state is 0, 1, 2, 3, …, N, and the remainder is in the other state.

Let us now look at a situation where the energy of the two states is not quite the same and the system is described by a ground state at zero energy and an excited state at energy, E. For distinguishable (i.e., classical) particles with $E = 0$, the particles will be equally divided between the two states. As E is increased the Maxwell–Boltzmann distribution shows that the occupancy of the excited state at a temperature T, decreases with increasing E as $\exp(-E/k_BT)$.

In the case of indistinguishable particles, there is no statistical tendency for the particles to be equally distributed between the two states when $E = 0$. Thus, when an energy difference exists between the two states, there is a greater probability (than in the case for distinguishable particles) for particles to transition into the lower-energy state. It is this fundamental property of indistinguishability in quantum mechanics, combined with the fact that bosons do not have to obey the Pauli principle, that gives rise to the formation of a Bose–Einstein condensate in which virtually all of the particles of a system at low temperature occupy the ground state.

7.4 What is a boson and what is not a boson

In order to study the behavior of a system of bosons, it is necessary to collect a number of bosons of the same type together and to confine these particles in some way that will enable us to investigate their properties. This is difficult with bosonic elementary particles, e.g. photons, mesons etc. so we need to look for a type of boson that is stable and can be collected together and confined. This means we need to look at atoms which are bosons, i.e. those that have an integer spin. Although, in principle, it is simple to determine whether a particular atom is a boson, it is important to fully understand how the spin of an atom arises.

The spin of an atom is the result of the spin of the nucleus and the spin associated with the atomic electrons. We begin with the nucleus. The nucleus is comprised of N neutrons and Z protons, all of which are spin-1/2 fermions. The total spin of the nucleus, $\boldsymbol{I}$, is the vector sum of the spins of all the neutrons, $\boldsymbol{J}_n$, and all the protons, $\boldsymbol{J}_p$:

$$\boldsymbol{I} = \boldsymbol{J}_n + \boldsymbol{J}_p.$$

The spin of the neutrons is the vector sum of the spins of the individual neutrons, $\boldsymbol{j}_{ni}$:

$$\boldsymbol{J}_n = \sum_i^N \boldsymbol{j}_{ni}$$

and similarly, for the protons:

$$\boldsymbol{J}_{\mathrm{p}} = \sum_{i}^{Z} \boldsymbol{j}_{\mathrm{p}i}.$$

The total spin of a neutron or proton is the vector sum of its orbital angular momentum, $\boldsymbol{l}$, and its intrinsic spin, $\boldsymbol{s}$:

$$\boldsymbol{j} = \boldsymbol{l} + \boldsymbol{s}.$$

For neutrons and protons in the nucleus, there is a strong coupling between the intrinsic spin and the orbital angular momentum (i.e., the strong spin–orbit coupling), leading to the relationship

$$j = l \pm \frac{1}{2}, \tag{7.2}$$

where 1/2 is the intrinsic spin of the neutron and proton. Since l is an integer (i.e. 0 for an s-state, 1 for a p-state, etc.) then for each neutron and each proton the total spin, j, calculated using equation (7.2) is always a half-integer. Although the vector relationships of the spins of the individual neutrons and protons are not necessarily easy to determine, it is clear from the above discussion that a bound system of neutrons and protons has an integer spin if the total number of particles is even, but such a system has a half-integer spin if the total number of particles is odd.

We now need to add the spin associated with the atomic electrons to the spin associated with the nucleus. Analogous to the discussion above, the total spin of the atomic electrons is the vector sum of the orbital angular momenta and the intrinsic spins of all the electrons. Again, while it is not straightforward to understand the vector relationships of all these spins, it follows from the discussion above that if a system consists of an even number of electrons, then its net spin is an integer, but if it consists of an odd number of electrons, then its net spin is a half-integer.

It should be noted that the spin quantum numbers represent a quantized angular momentum in units of $\hbar$, which has units in the SI system of J s. Thus, although we may view electrons as quite different from neutrons and protons, in terms of mass and spatial distribution in the atom, it is appropriate to merely (vectorially) add the spin of the nucleus and the spin of the electrons to obtain the total spin of the atom. Thus, we again come to the simple conclusion that while the vector relationship of the electron spin and the nuclear spin may not be known, if the total number of electrons, neutrons, and protons in an atom is even, then the atom is a boson. Conversely, if the total number of electrons, neutrons, and protons is odd, then the atom is a fermion.

We now have a simple prescription for determining whether a given atom is a boson or a fermion, and this enables us to choose atoms that are bosons for a Bose–Einstein condensate experiment. Since the number of electrons in a neutral atom must be equal to the number of protons, it is obvious that the number of electrons plus the number of protons must be an even number. This gives us the simple criterion for determining whether an atom is a boson: all neutral atoms with an even number of neutrons are bosons. This means that different isotopes of the same element, which differ in their numbers of neutrons, are bosons when the neutron number is even and fermions when the neutron number is odd.

We can now look at some specific examples. A ^{4}He atom which consists of two neutrons, two protons, and two electrons, i.e. one of the simplest atoms, is also a boson. While a Bose–Einstein condensate has not been observed in ^{4}He, it does exhibit another interesting property, superfluidity, which is related to the fact that its atoms are bosons. The first true Bose–Einstein condensate was observed in 1995 by Eric Cornell and Carl Wieman for dilute ^{87}Rb atoms. A few months later, Wolfgang Ketterle observed a Bose–Einstein condensate for ^{23}Na. In more recent years, Bose–Einstein condensates have been observed for dilute gases of a number of other atoms, including ^{7}Li, ^{23}Na, ^{39}K, ^{40}Ca, ^{41}K, ^{52}Cr, ^{84}Sr, ^{85}Rb, ^{86}Sr, ^{87}Rb, ^{88}Sr, ^{133}Cs, ^{164}Dy and ^{168}Er and ^{174}Yb. While roughly half of all nuclear species have an even number of neutrons in their nucleus, it is clear from the list of isotopes in which this phenomenon has been observed that the vast majority belong to specific portions of the periodic table, specifically group 1 (alkali metals), group 2 (alkaline earth metals), and the lanthanides. The reasons for this have to do with the interactions between atoms, as discussed in further detail in the next chapter.

References and suggestions for further reading

Cornell E A and Wieman C E 1998 The Bose–Einstein condensate *Sci. Am.* March 40–5

Kittel C 1969 *Thermal Physics* (Hoboken, NJ: Wiley)

Maschen 2011 Vector model of orbital angular momentum https://commons.wikimedia.org/wiki/File:Vector_model_of_orbital_angular_momentum.svg

Pitaevskii L P and Stringari S 2003 *Bose–Einstein Condensation* (Oxford: Clarendon) https://global.oup.com/academic/product/bose-einstein-condensation-9780198507192

Reif F 1965 *Fundamentals of Statistical and Thermal Physics* (New York: McGraw-Hill)

Chapter 8

Bose–Einstein condensates

8.1 Introduction

After the process of cooling and trapping bosons as described in chapters 5 and 6 is completed, a collection of typically a few thousand to a few million atoms at a temperature below 100 nK is trapped in a volume which is less than one millimeter on a side. The creation of such a system opens up the possibility of understanding the behavior of ultracold atoms and the possibility of making use of their unusual properties. This chapter reviews the properties of Bose–Einstein condensates and considers the methods by which their existence can be confirmed. In addition, it presents the results of some experiments that used novel approaches to investigate this phenomenon.

8.2 Bose–Einstein condensation

The unusual behavior of a Bose–Einstein condensate occurs for a system of bosons which is at a temperature that is low enough for virtually all of the particles to be in their ground state. It is in this state that, under the right conditions, the collection of bosons exhibits large-scale collective quantum mechanical behavior. This situation occurs when the collection of bosons is dilute enough that there is minimal interaction between individual particles and the de Broglie wavelength is large enough that their wave functions overlap. Alkali metals which have integer spin are some of the most suitable atoms for use in the creation of Bose–Einstein condensates. Alkali metals have filled electronic shells plus one additional loosely bound s-state electron. In a dilute gas, there is a weak repulsive interaction between atoms because of this outer electron, and this enables the atoms to exist in their bosonic ground state with a minimum of physical interaction.

doi:10.1088/978-0-7503-5482-0ch8

An approximate value for the temperature at which Bose–Einstein condensation occurs can be found by the following simple derivation. The de Broglie wavelength of a particle with linear momentum, p, is

$$\lambda = \frac{h}{p},$$

where h is the Planck constant. For non-relativistic particles of mass m, the energy is given by

$$E = \frac{p^2}{2m}.$$

Therefore,

$$\lambda = \frac{h}{\sqrt{2mE}}. \tag{8.1}$$

The thermal energy of a particle in three dimensions at a temperature, T, is

$$E = \frac{3}{2}k_B T. \tag{8.2}$$

Combining equations (8.1) and (8.2) gives

$$\lambda = \frac{h}{\sqrt{3mk_B T}}. \tag{8.3}$$

For the situation described above, the Bose–Einstein condensation temperature, T_{BE}, is the temperature at which the spatial extent of the de Broglie wave becomes greater than the average distance between the atoms. For a gas with a number density of atoms n (measured in atoms per unit volume), the average distance between atoms is $n^{-1/3}$, so that the Bose–Einstein condition is met in terms of equation (8.3) when

$$\frac{1}{n^{1/3}} \leqslant \frac{h}{\sqrt{3mk_B T_{BE}}}.$$

Solving this expression for temperature gives

$$T_{BE} \leqslant \frac{h^2 n^{2/3}}{3mk_B}. \tag{8.4}$$

A more detailed calculation actually gives a value for the Bose–Einstein temperature which is about a factor of three higher than that given by equation (8.4), but the above simple analysis provides some physical insight into the meaning of this behavior. In order to ensure that the atoms in the condensate do not interact with one another, they cannot be too close together. This puts a practical limit on the density, n, and subsequently imposes limits on how high T_{BE} can be. Typical experiments require temperatures in the nano-Kelvin range; thus, the experimental techniques described in chapter 5 for cooling along with the techniques described in chapter 6 for confining a dilute collection of atoms are most commonly necessary for the observation of a Bose–Einstein condensate.

8.3 Creating and identifying a Bose–Einstein condensate

The most straightforward method of confirming that a collection of bosons has formed a Bose–Einstein condensate is to show that its velocity distribution is characteristic of such a condensate and not of a classical Maxwell–Boltzmann distributed gas. Measuring the velocity distribution of the atoms is relatively straightforward. The magnetic field producing the trap is turned off, allowing the atoms to move freely. Two things then happen to the movement of the atoms. First, the atoms are subject to gravitational force and start to fall. Second, since the atoms are no longer trapped by the magnetic field, their spatial distribution starts to spread. These changes in atomic motion can be observed using a laser. The laser is tuned to the resonant frequency of the atomic excited state. There is, therefore, a high probability of resonant photon absorption by the atoms and the accompanying population of the excited state. There is, of course, re-emission of these absorbed photons, but this is isotropic, as discussed above. As a result of this absorption, there is a large decrease in the intensity of the laser beam in the forward direction. The amount of light that is transmitted through the gas is inversely proportional to the density of the gas. The light that is transmitted through the gas can be imaged using a charge coupled device (CCD), which is the type of sensor used in digital cameras. Changes in the spatial distribution of atoms can be observed by monitoring the changes in the image recorded by the CCD, and these can be related to the distribution of velocities in the two dimensions orthogonal to the propagation direction of the laser.

Figure 8.1 shows the results of two-dimensional velocity distribution measurements for a gas of ^{87}Rb atoms undergoing Bose–Einstein condensation. At the left, the temperature is greater than the Bose–Einstein condensation temperature for the gas. The velocity distribution is characterized by the broad Maxwell–Boltzmann distribution characteristic of a classical gas. In the central image, the gas has been cooled to a temperature lower than the condensation temperature, and a very sharp peak appears in the center of the velocity distribution due to the atoms that have condensed into a Bose–Einstein condensate. In the image on the right, the temperature has been lowered by removing the higher-velocity atoms using forced evaporative cooling, and the resulting collection of atoms has formed a nearly pure Bose–Einstein condensate.

Figure 8.2 shows another example of a Bose–Einstein condensate. In this case the bosons are $^{40}K_2$ molecules. A ^{40}K atom has 21 neutrons and 19 protons in the nucleus and 19 atomic electrons. Thus, a neutral ^{40}K atom has a total of 59 fermions and is, therefore, a fermion itself. The $^{40}K_2$ molecule consists of two ^{40}K atoms and is therefore a boson. The results shown in the figure illustrate, from right to left, the formation of a Bose–Einstein condensate of $^{40}K_2$ molecules with decreasing temperature.

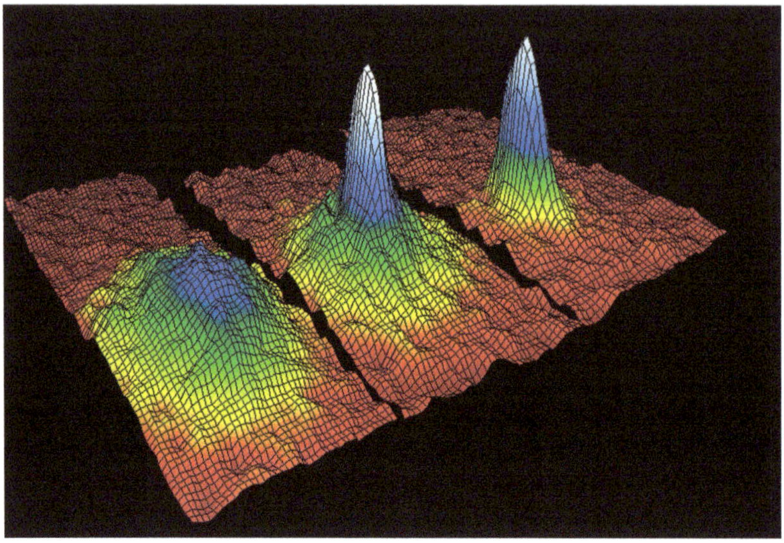

Figure 8.1. The formation of a Bose–Einstein condensate as a function of temperature in a collection of ^{87}Rb atoms. (left) Before the formation of the Bose–Einstein condensate, showing a Maxwell–Boltzmann velocity distribution. (center) Just after the formation of the condensate, showing the sharp Bose–Einstein peak in the center. (right) A nearly pure Bose–Einstein condensate. Reproduced from NIST/JILA/CU-Boulder (2006). Image stated to be in the public domain.

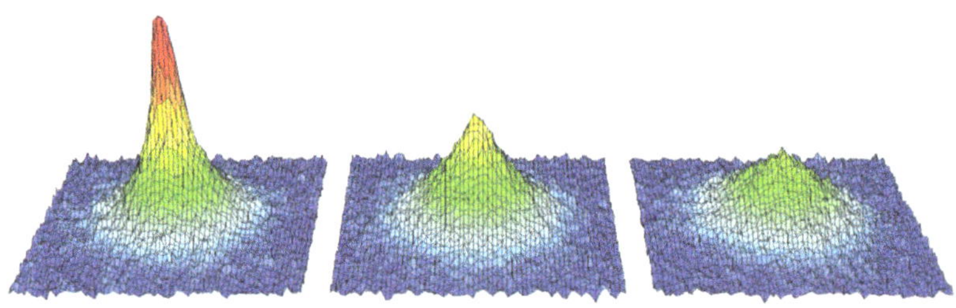

Figure 8.2. Formation of a Bose–Einstein condensate by bosonic $^{40}K_2$ molecules, with decreasing temperature from right to left. Reprinted (figure) with permission from Regal *et al* (2004), Copyright (2004) by the American Physical Society.

Another interesting and important feature that is seen in figures 8.1 and 8.2 relates to the shape of a cross section of the peak in a plane parallel to the velocity axes. It can be observed that the high-temperature Maxwell–Boltzmann distribution has a circular cross section. This is because the Maxwell–Boltzmann velocity distribution is isotropic in three-dimensional space. It can be seen, however, that the sharp peak that arises from the Bose–Einstein condensate has a cross section that is elliptical, not circular. As can be noted, the velocities that are shown in the graphs are the velocities in the x- and z-directions (not x and y) and this is the reason for the elliptical shape. The velocity distribution of the Bose–Einstein condensate is not isotropic. Rather, it assumes a shape that is characteristic of the spatial geometry of the magnetic field trap. This is because it is, in fact, the magnetic field distribution, not the position in real space, that defines the Bose–Einstein condensate. Based on the orientation of the magnetic coils, the magnetic field has axial symmetry in the x–y plane but differs in the z-direction. As a consequence of the three-dimensional characteristics of the magnetic field lines, the x and z dependences of the velocity are different, leading to the anisotropy seen in the figure. In fact, some experiments intentionally alter the anisotropy of the magnetic field profile in the z-direction in order to accentuate the anisotropy of the Bose–Einstein velocity distribution as a convenient means of distinguishing between the isotropic Maxwell–Boltzmann distribution and the anisotropic Bose–Einstein distribution.

8.4 Microgravity Bose–Einstein condensate experiments

Quantum effects are typically microscopic properties of physical systems in which the effects of gravity are negligible. Bose–Einstein condensates are a manifestation of quantum mechanical effects that govern the behavior of macroscopic collections of atoms. As such, they form an example of a situation in which quantum phenomena are influenced by weak external forces such as gravity. Specifically, the force of gravity causes atoms to fall out of the atomic traps that are created by the combined action of lasers and magnetic fields. In order to access systems at lower temperatures, weaker trapping fields are required, leading to shorter observation times before the effects of gravity remove atoms from the trap. The possibility of performing Bose–Einstein condensate experiments in microgravity aboard the International Space Station (ISS) provides a means of overcoming the limitations of experiments performed on Earth. In 2018, the Cold Atom Lab (CAL) was established on board the ISS and has utilized the microgravity environment to better understand the properties of ultracold matter (Aveline *et al* 2020).

The significance of Bose–Einstein condensate experiments conducted in microgravity is demonstrated by the results for a ^{87}Rb condensate as shown in figure 8.3. The microgravity condensate exhibits a diffuse halo around the central concentration of atoms which is not seen in the Earth-based condensate. This behavior of the Bose Einstein condensate can be explained on the basis of the magnetic interactions between the bosons and the applied magnetic field. Although, as noted in chapter 6, the magnetic moment associated with the atomic electrons is responsible for the interactions that determine the trapping of the atoms in the magneto-optical trap, the angular momentum associated with the nucleus is of importance in certain circumstances. The total angular momentum of the atom, $\boldsymbol{F}$, is expressed in terms of the nuclear spin, $\boldsymbol{I}$, and the net orbital angular momentum of the electrons, $\boldsymbol{J}$, as

$$\boldsymbol{F} = \boldsymbol{I} + \boldsymbol{J}.$$

The z component of $\boldsymbol{F}$ can take the values

$$m_F = -F, -F + 1 \cdots F - 1, F.$$

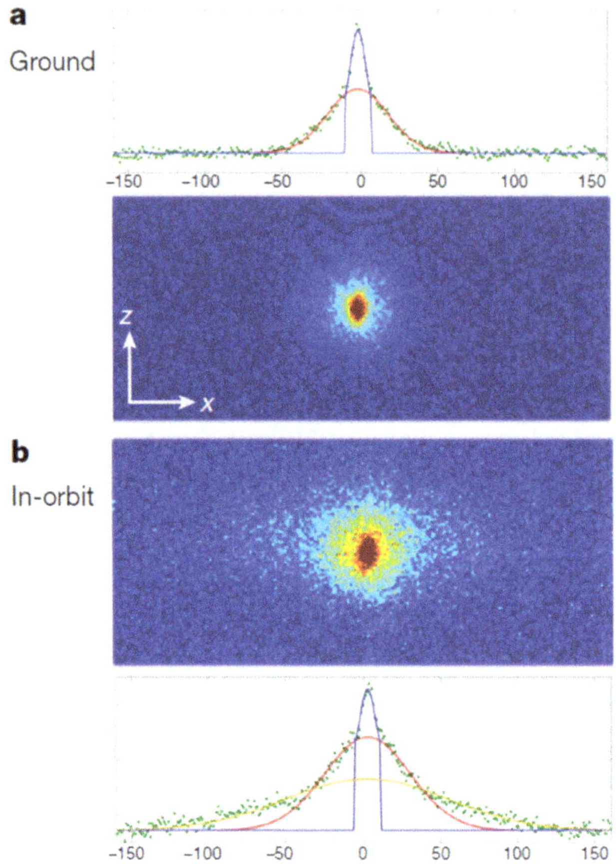

Figure 8.3. Observation of a ^{87}Rb Bose–Einstein condensate (a) on Earth and (b) in microgravity on board the ISS as shown in a false-color absorption image. The scale on the horizontal axis is in pixels. Aveline *et al* (2020) (Copyright © 2020, The Author(s), under exclusive licence to Springer Nature Limited). With permission of Springer.

The ^{87}Rb atoms which interact with the applied magnetic field and are, therefore, trapped in the magneto-optical trap, are in the state

$$|\, F, m_F > = |\, 2{,}2 >.$$

Atoms in the $|\, 2{,}0\rangle$ state are magnetically insensitive and are not trapped in the magneto-optical trap. Experiments have shown that of the order of half of the ^{87}Rb atoms are typically in this state. On Earth, the dominant force that determines the behavior of these atoms is gravity, and they quickly fall out of the region of the trap. In microgravity, the dominant force affecting these atoms is the result of the quadratic Zeeman effect (Jenkins and Segrè 1939), which is proportional to the square of the applied magnetic field. It is the significance of the quadratic Zeeman effect in the absence of gravity that is responsible for the existence of the halo around the trapped Bose–Einstein condensate in figure 8.3. Further studies using magnetic field gradients have confirmed that this component of the atomic distribution is associated with ^{87}Rb atoms in the $|2,0\rangle$ state.

The above example of results from Bose–Einstein condensate experiments conducted in microgravity demonstrates the usefulness of this approach. Gravity, which normally has no significant effect on experiments dealing with the quantum mechanical nature of materials, can be an important factor in the behavior of the delicate structures that are created in systems of ultracold atoms. Eliminating the effects of gravity can therefore provide insights into important fundamental physics that cannot be observed in a normal laboratory environment.

The Cold Atom Lab on board the International Space station is an ongoing experiment that has the options of investigating ^{87}Rb and ^{41}K condensates. It is operated remotely from the Jet Propulsion Laboratory in Pasadena, California, and provides a platform for future experiments involving the properties of ultracold atoms in a microgravity environment.

8.5 Quasiparticle Bose–Einstein condensates

The Bose–Einstein condensate experiments described thus far all require ultralow temperatures. An approximate value for the necessary temperature is given above in equation (8.4). For a given boson system, this expression provides the maximum temperature at which condensates can be formed. This is because there is a limit on the number density of bosons in the system in order to ensure that there are minimal interactions between the individual particles. This typically corresponds to a gas density of the order of 10^{-5} smaller than the density of air.

The first observation of a Bose–Einstein condensate was for a system of ^{87}Rb atoms (Anderson *et al* 1995) for which the condensation temperature was measured to be 170 nK. Since that time, traditional Bose–Einstein condensates have been observed in ultracold collections of a variety of atomic species ranging (in mass) from ^{7}Li to ^{174}Yb. From the mass dependence in equation (8.4), it can be seen that the transition temperatures of other systems fall within an order of magnitude or so of the value for ^{87}Rb. The following discussion considers some of the possibilities that have been explored in systems in which the Bose–Einstein transition temperature may be much higher. While laser cooling and trapping, as discussed previously, is not necessary for these experiments, lasers still play an important role in the observation of Bose–Einstein condensates, as described below.

8.5.1 Bose–Einstein condensation of excitons

The possibility of observing Bose–Einstein condensates at temperatures accessible by standard cryogenic technologies (e.g. >1 K), or even at room temperature, requires a system in which the boson mass is many orders of magnitude smaller than that of a bosonic atom. One step toward achieving transition temperatures above about 1 K would be to use systems of bosons with a mass comparable to an electron. This would place the transition temperature in the range of tens of mK. The first experimental investigations that took this approach studied the condensation of excitons. An exciton is a quasiparticle consisting of a bound electron–hole pair and,

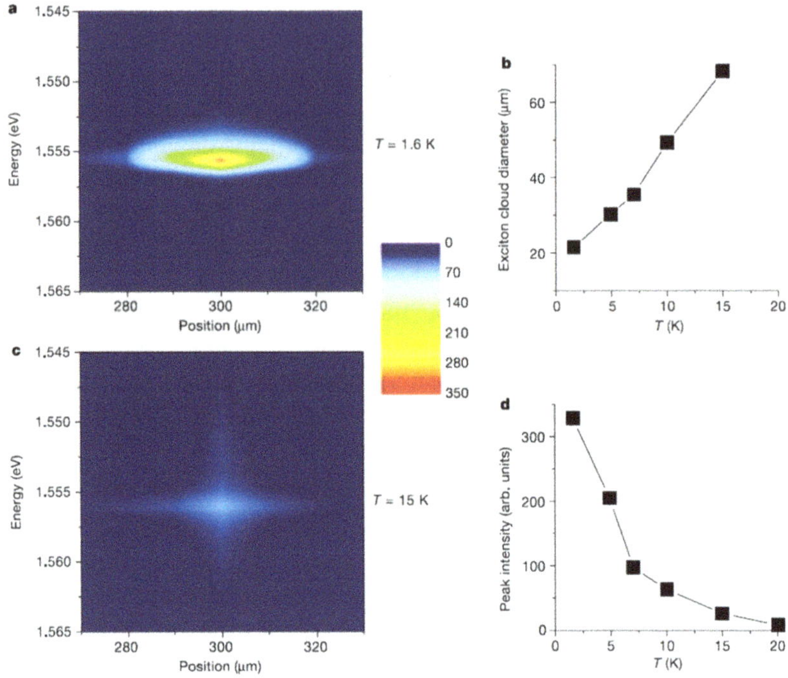

Figure 8.4. Observations of an exciton cloud as a function of temperature at (a) 1.6 K and (c) 15 K. The temperature dependence of the diameter of the exciton cloud and the peak photoluminescent intensity are shown in (b) and (d), respectively. A He–Ne laser was used to excite the sample. Butov *et al* (2002) (Copyright © 2002, Nature Publishing Group). With permission of Springer.

as such, is a boson that comprises two fermions. Excitons most commonly exist in semiconductors, where their overall mass is the effective mass, m_{eff}, of the bound pair and can be much smaller than the mass of a free electron.

Butov *et al* (2002) studied this phenomenon using photoluminescence measurements of a GaAs quantum well. In such a structure, excitons exist in quasi two-dimensional traps. Traps are formed by naturally occurring fluctuations in the dimensions of the semiconductor heterostructure. The Bose–Einstein transition temperature for a two-dimensional square potential may be calculated to be (Butov *et al* 2002)

$$T_{\text{BE}} = \frac{h^2 n}{2\pi m_{\text{eff}} g k_{\text{B}}} \cdot \frac{1}{\ln\left[\frac{nS}{g}\right]}, \tag{8.5}$$

where n is the exciton density, S is the area of the two-dimensional trap, and g is the spin degeneracy of the exciton state. In the case of the GaAs quantum well, $m_{\text{eff}} = 0.21 m_{\text{e}}$ and $g = 4$, leading to a transition temperature on the order of 1 K according to equation (8.5).

Figure 8.4 shows the results of a photoluminescence study conducted using a He–Ne laser. The study examined excitons in a quasi two-dimensional GaAs quantum well. The temperature-dependent data in the figure show several significant features. At higher temperatures, the excitons are distributed over the more spatially extended higher-energy states in the trap; as the temperature is lowered, the excitons condense down into the more spatially restricted lower-energy states. This narrowing of the spatial distribution of excitons is accompanied by an increase in the peak intensity of the distribution. These features are consistent with the behavior observed in dilute atomic Bose–Einstein condensation experiments and indicate that this condensation is occurring in excitons trapped in the GaAs quantum well. The results also indicate that the condensation, although possibly not complete at the lowest measurement temperature, is occurring at a temperature that is consistent with predictions for exciton condensation, and this provides a means of observing the effects of Bose–Einstein condensation at temperatures that are accessible using traditional cryogenic technology.

8.5.2 Bose–Einstein condensation of polaritons

Polaritons are quasiparticles consisting of an electromagnetic wave (photon) coupled to an excitation which carries an electric or magnetic dipole moment. In the present context, the polariton is a photon coupled to an exciton, namely, a bound electron–hole pair, as described above. Since the polariton is a coupling of two bosons, it is, itself, a boson. Polaritons can be created in an optical microcavity in a layered semiconductor, as illustrated in figure 8.5(a). The optical microcavity is

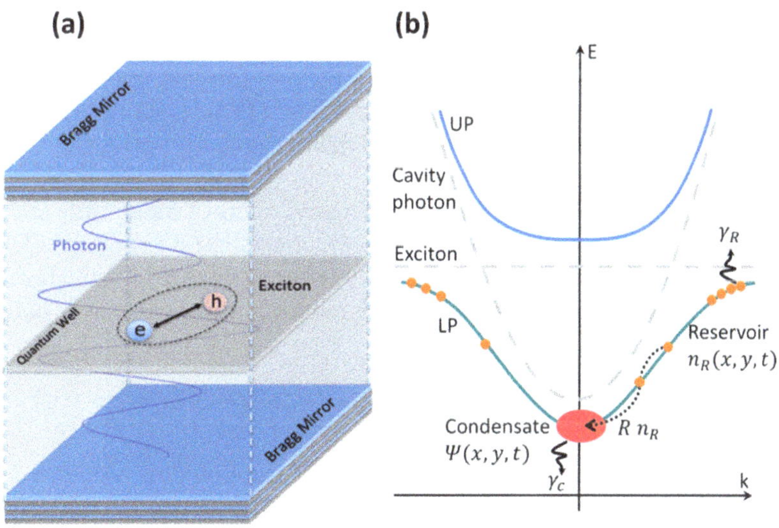

Figure 8.5. (a) A diagram of a microcavity created in a layered semiconductor device as described in the text. (b) The dispersion relation, showing the upper polariton (UP) and lower polariton (LP) branches. The exciton and photon energies are illustrated as a function of the wave vector. Reprinted with permission from Gargoubi *et al* (2016). Copyright 2016 by the American Physical Society.

Figure 8.6. The formation of a polariton Bose–Einstein condensate at $T = 5$ K as a function of excitation power density. From left to right, the power densities are 0.919 kW·cm^{-2}, 1.671 kW·cm^{-2}, and 1.905 kW·cm^{-2}, and the threshold power for condensation is 1.671 kW·cm^{-2}. Kasprzak *et al* (2006) Copyright © 2006, Nature Publishing Group. Reproduced with permission from Springer Nature.

created by depositing layers of semiconductor that have different indices of refraction. Photons provided by an external excitation laser resonate inside the cavity, a situation that is analogous to the resonance condition established in a conventional laser. The microcavity incorporates an active layer where quantum wells are formed, and excitons are created from electron–hole pairs. Polaritons are formed by the interaction between resonant photons inside the microcavity and excitons in the quantum-well layer, as seen in the figure. Under suitable conditions and at sufficiently low temperatures, polaritons (bosons) condense at the bottom of the dispersion relation, as shown in figure 8.5(b).

The first experimental evidence for the Bose–Einstein condensation of polaritons was presented by Kasprzak *et al* (2006). In this work, a CdTe/CdMgTe microcavity was grown by molecular-beam epitaxy and excitation was accomplished using a continuous-wave Ti:Al_2O_3 laser (see section 5.4). Figure 8.6 shows some results from this study that demonstrate the formation of a Bose–Einstein condensate. The far-field polariton emission pattern at a fixed temperature (5 K) is shown as a function of the excitation power. The formation of a sharp intense central peak characteristic of a Bose–Einstein condensate can be observed.

As the effective mass of polaritons can be many orders of magnitude less than the free electron mass, there is the possibility of creating Bose–Einstein condensates at even higher temperatures; recently, evidence has been presented for the formation of Bose–Einstein condensates of polaritons in microcavities near room temperature (Plumhof *et al* 2013, Das *et al* 2013).

8.6 Why is it useful?

Although the Bose–Einstein condensate does not have obvious commercial applications, it has been important in advancing our fundamental understanding of quantum systems, wave coherence, and interference effects. Not only does the study of Bose–Einstein condensates validate the theoretical framework of the statistics of particles and the properties of de Broglie waves, but it provides a convenient platform for investigations of many properties of condensed systems. A 'lattice' can be constructed from the interference patterns of multiple laser beams, as described in the next chapter. Since the atoms in a Bose–Einstein condensate can be manipulated by means of laser beams, this periodic optical lattice influences the atoms of a Bose–Einstein condensate in a way that is analogous to the way in which crystalline lattices influence electrons in normal solids. The optical lattice, however, can be manipulated much more easily and in ways that are not possible for a crystalline lattice. For example, the lattice periodicity, interactions, and dimensionality can be controlled by the experimental parameters related to the properties of the laser beams and the Bose–Einstein condensate. This ability allows for the convenient testing of solid-state models and the investigation of phenomena that are not easily accessible by conventional solid-state experiments.

Bose–Einstein condensates are, by their very nature, fragile, and the use of their properties in practical devices poses substantial challenges. However, their ability to respond to very small perturbations may make them ideal for certain uses. It has been proposed that gravitational waves, as predicted by Einstein's theory of general relativity, may be detected by means of a Bose–Einstein condensate. Gravitational waves are expected to generate phonons in a Bose–Einstein condensate, and the influence of these phonons on the properties of the condensate would be observable. This would mean that interference effects in the matter waves in the condensate could be used to produce a tabletop-sized experiment that could replace the kilometer-long optical interferometers currently used to observe gravitational waves.

The very stable wave function associated with a Bose–Einstein condensate could form the basis for a highly accurate time standard that could improve upon current technologies which utilize the frequency of electronic transitions in atoms as a basis for measuring time. Other potentially more commercially viable applications include matter holograms, which use the interference of matter waves in the same way that optical holograms use light; quantum computers, which use the quantum state of atoms in a condensate to store information; and nanolithography, in which atomic-scale features could be created on electronic devices using precision-controlled condensates. Some of these possibilities are discussed further in the next chapter.

References and suggestions for further reading

Anderson M H, Ensher J R, Matthews M R, Wieman C E and Cornell E A 1995 Observation of Bose-Einstein Condensation in a Dilute Atomic Vapor *Science* **269** 198–201

Aveline D *et al* 2020 Observation of Bose–Einstein condensates in an earth-orbiting research lab *Nature* **582** 193–7

Butov L V, Lai C W, Ivanov A L, Gossard A C and Chemla D S 2002 Towards Bose–Einstein condensation of excitons in potential traps *Nature* **417** 47–52

Cornell E A and Wieman C E 1998 The Bose–Einstein condensate *Sci. Am.* March 40–5

Das A, Bhattacharya P, Heo J, Banerjee A and Guo W 2013**110** 2735–40

Davis K B, Mewes M O, Andrews M R, van Druten N J, Durfee D S, Kurn D M and Ketterle W 1995 Bose–Einstein condensation in a gas of sodium atoms *Phys. Rev. Lett.* **75** 3969–73

Gargoubi H, Guillet T, Jaziri S, Balti J and Guizal B 2016 Polariton condensation threshold investigation through the numerical resolution of the generalized Gross–Pitaevskii equation *Phys. Rev.* E **94** 043310

Griffin A, Snoke D W and Stringari S (ed) 1995 *Bose–Einstein Condensation* (Cambridge: Cambridge University Press)

Jenkins F A and Segrè E 1939 The quadratic Zeeman effect *Phys. Rev.* **55** 52–8

Junttila M L and Stahlberg B 1990 Laser wavelength measurement with a Fourier transform wavemeter *Appl. Opt.* **29** 3510–6

Kasprzak J *et al* 2006 Bose–Einstein condensation of exciton polaritons *Nature* **443** 409–14

NIST/JILA/CU-Boulder 2006 3-D of Atom Condensing https://www.nist.gov/image/quantumphysicsboseeinsteincondensatejpg

Pethick C J and Smith H 2001 *Bose–Einstein Condensation in Dilute Gases* (Cambridge: Cambridge University Press)

Pitaevskii L P and Stringari S 2003 *Bose–Einstein Condensation* (Oxford: Clarendon) https://global.oup.com/academic/product/bose-einstein-condensation-9780198507192

Plumhof J D, Stöferle T, Mai L, Scherf U and Mahrt R F 2013 Room-temperature Bose–Einstein condensation of cavity exciton–polaritons in a polymer *Nat. Mater.* **13** 247–52

Regal C A, Greiner M and Jin D S 2004 Observation of resonance condensation of fermionic atom pairs *Phys. Rev. Lett.* **92** 040403

Chapter 9

Other applications of laser cooling and trapping

9.1 Introduction

While the study of Bose–Einstein condensates was one of the earliest uses of laser cooling and trapping technology, there are a number of other possible applications of this technology. Some of these are important research tools for the study of fundamental physical phenomena, while others may have future practical applications. A few of these possible applications of laser cooling and trapping are discussed in this chapter.

9.2 Atomic fountains

Accurate timekeeping has a variety of important practical applications. These include the accurate coordination of global times with changes in the Earth's rotational period (achieved by the introduction of leap seconds) and accurate positioning using global positioning systems (GPSs). Applications also include experimental tests of special and general relativity. In the 1950s, atomic clocks were developed, and these provided considerably improved accuracy compared to the mechanical timepieces available at the time. Most atomic clocks have utilized the energy difference between two hyperfine ground states of cesium-133 (^{133}Cs) atoms, and this transition has been used to define the SI unit of time, the second (Taylor 2001).

The atomic oscillator of an atomic clock is characterized by the quality factor, Q, defined as

$$Q = \frac{\nu_0}{\Delta\nu}, \tag{9.1}$$

where ν_0 is the central frequency of the transition and $\Delta\nu$ is the transition width. A major contribution to the transition width is the Doppler broadening that results from thermal motion. The clock's performance is described by the fractional instability expressed by the Allan standard deviation (Allan 1966). This is given by (Itano *et al* 1993)

$$\sigma_y \approx \frac{1}{Q\sqrt{n\tau}}, \tag{9.2}$$

where n is the number of atoms measured per unit time and τ is the total measurement time. Maximizing clock performance therefore requires maximization of the quality factor as well as maximization of the number of atoms sampled and the total measurement time. In addition, it is important to minimize the clock's sensitivity to external perturbations.

The atomic fountain forms the basis of an atomic clock design; it was first proposed by the American physicist Jerrold Zacharias (1905–86) in the 1950s (see Wynands and Weyers 2005). Early attempts to create useful atomic fountains were not successful because of the atom–atom interactions that resulted from thermal energy. As described previously in this book, the development of technologies utilizing laser cooling provided a means of overcoming this problem.

Figure 9.1 shows the principle of operation of an atomic fountain clock. In figure 9.1(a), an optical or magneto-optical trap cools and traps appropriate atoms using lasers tuned to a frequency ν. In figure 9.1(b), the upward-directed laser is detuned to a frequency of $\nu + \delta\nu$, and the downward-directed laser is detuned to a frequency of $\nu - \delta\nu$. This launches the atoms upward at a velocity, v, given by

$$v = \frac{\delta\nu}{\nu}c,$$

where c is the speed of light. In figure 9.1(c), the atoms travel upward in a parabolic path through a microwave cavity. In figure 9.1(d), the atoms fall downward under the force of gravity, back through the microwave cavity and past the detection laser.

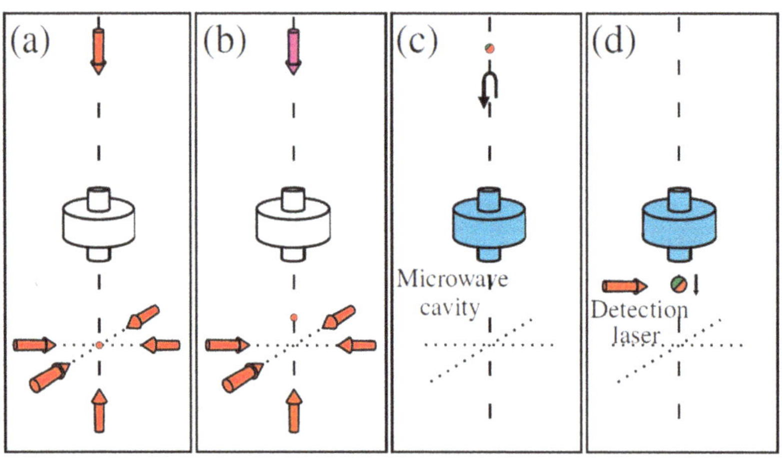

Figure 9.1. The operational principle of an atomic fountain clock. (a) The trapping of a collection of cold atoms in an optical trap formed by three orthogonal pairs of counterpropagating sets of lasers. (b) The detuning of vertical lasers to impart a vertical velocity to the atoms. (c) The atoms follow a parabolic trajectory and pass through a microwave cavity while traveling upward and again while traveling downward. (d) The population distribution is observed using fluorescence induced by the detection laser. Reprinted from Wynands and Weyers (2005) © IOP Publishing. Reproduced with permission. All rights reserved.

Figure 9.2 shows a simplified diagram of an atomic fountain clock. In this design, the three sets of orthogonal counterpropagating lasers are aligned so that two sets of lasers lie in a horizontal plane and the third orthogonal set is oriented vertically. This means that the vertical laser beam travels through the hole in the microwave cavity through which the atoms also travel. This configuration limits the diameter of the laser beam to the diameter of the hole in the cavity, which is typically around

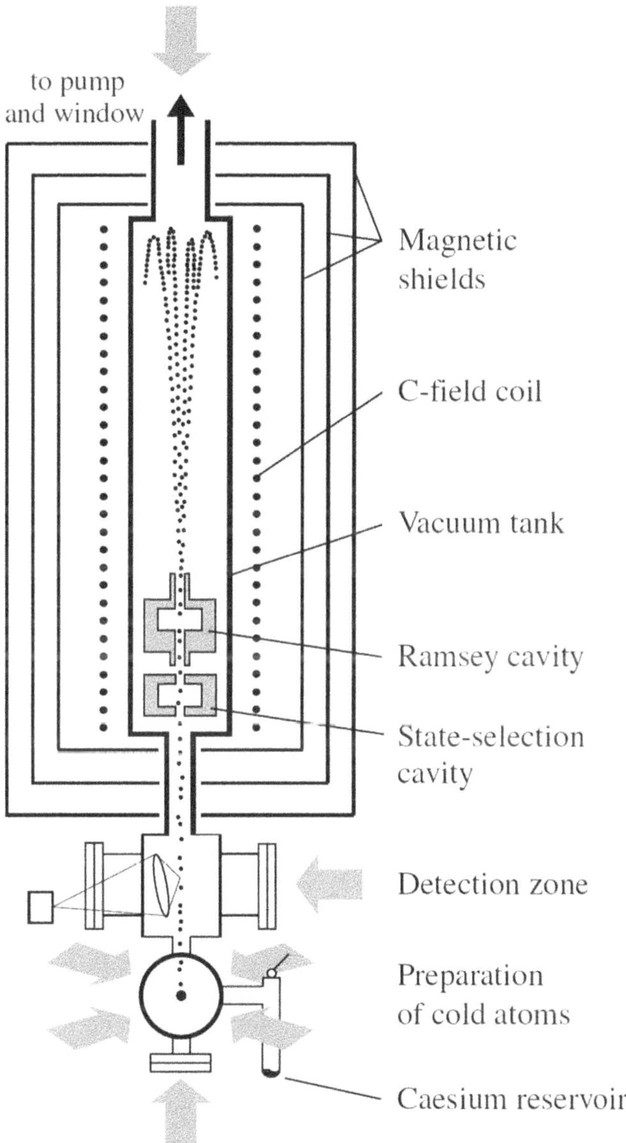

Figure 9.2. A diagram of a simple atomic fountain clock. The height of the fountain is typically around 50 cm. Reprinted from Wynands and Weyers (2005) © IOP Publishing. Reproduced with permission. All rights reserved.

1–2 cm. An alternative geometry that does not have this constraint is the so-called (1,1,1) geometry. The laser geometry in figure 9.2 can be viewed by imagining the beams as being perpendicular to the faces of a cube whose *z*-direction is perpendicular to a cube face that is in the horizontal plane. The (1,1,1) geometry is achieved by rotating the cube so that the (1,1,1) direction of the cube lies in the *z*-direction, thereby eliminating interference between the laser and the atoms as they travel up and down their trajectory. Another feature of the design illustrated in figure 9.2 is that its operation is pulsed. This means that the atoms are launched in pulses so that the downward-traveling atoms do not interact with the upward-traveling atoms. An alternative arrangement can be constructed in which the atoms' trajectories are angled slightly from the vertical and separate microwave cavities are used for the upward-traveling and downward-traveling atoms.

Traditionally, the frequency associated with the hyperfine split of the ground state of ^{133}Cs is used as a time standard. Figure 9.3 shows an energy level diagram of the ground state and the first excited state of ^{133}Cs. Of particular importance is the

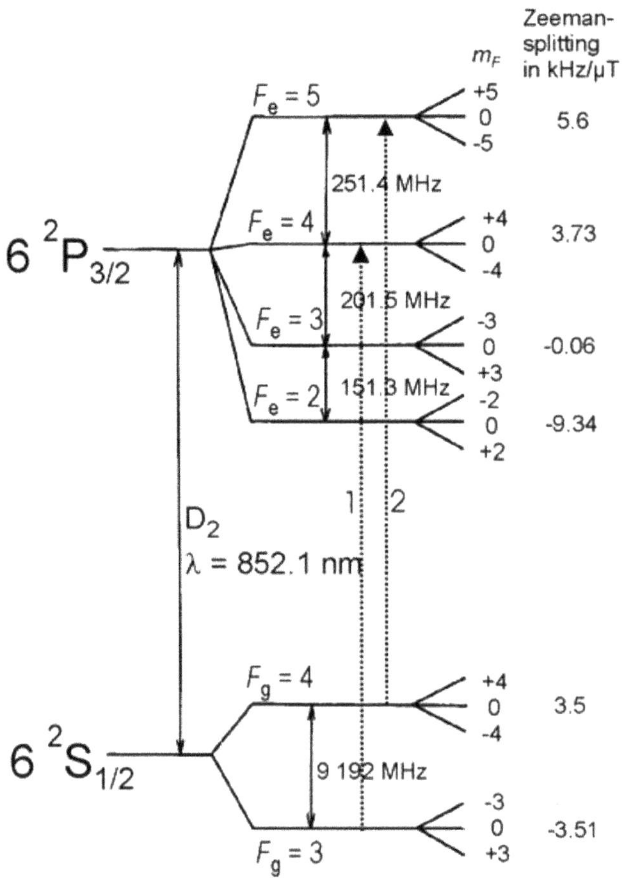

Figure 9.3. The ground-state and first excited-state energy levels of ^{133}Cs. Reprinted from Bauch (2002) © IOP Publishing. Reproduced with permission. All rights reserved.

hyperfine split of the $6^2S_{1/2}$ ground state into two levels with total spins of $F_g = 3$ and $F_g = 4$. This hyperfine split arises because of the interaction between the atomic electrons and a nonzero nuclear magnetic moment (I = 7/2 for ^{133}Cs). The total electron spin (1/2 for Cs) couples either parallel to the nuclear spin to give a total spin of $F = I + J = 7/2 + 1/2 = 4$ or antiparallel to the nuclear spin to give a total spin of $F = I - J = 7/2 - 1/2 = 3$, as shown in figure 9.4 (see section 8.4). Similarly, the 3/2 spin excited state has a total spin of 2, 3, 4, or 5.

The cooling lasers are red detuned relative to the energy associated with the $|F_g = 4\rangle \to |F_e = 5\rangle$ transition. A second laser at the energy of the $|F_g = 3\rangle \to |F_e = 4\rangle$ transition is superimposed on at least one of the cooling laser frequencies in order to repopulate the $|F_g = 4\rangle$ level, thereby allowing the continuation of the cooling process. After atoms are launched into the atomic fountain, they pass through the state-selection cavity, as shown in figure 9.2. This ensures that all of the atoms are in their $|F_g = 3\rangle$ ground state. The atomic beam then passes through the microwave cavity (Ramsey resonator) twice, once on the way up and once on the way down. The frequency of the microwave generator driving the cavity, ν, is tuned to the transition energy between the hyperfine split ground states of ^{133}Cs, E, and is given by

$$\nu = \frac{E}{h}. \tag{9.3}$$

Tuning the microwave generator to this resonant frequency maximizes the fraction of ground-state atoms that undergo the transition $|F_g = 3\rangle \to |F_g = 4\rangle$. The ratio of atoms in the $|F_g = 4\rangle$ state to the total number of atoms is measured in the detection zone in figure 9.2. This information is fed back into the microwave generator, as described in detail below, to adjust the frequency to ideally match the resonance

Figure 9.4. The NIST-F2 cesium fountain atomic clock at the U.S. Department of Commerce's National Institute of Standards and Technology (NIST) in Boulder, CO with NIST physicists Steve Jefferts (foreground) and Tom Heavner. Public domain, courtesy of NIST.

condition. The frequency of the ^{133}Cs ground-state transition in equation (9.3) is defined to be exactly 9 192 631 770 Hz. Therefore, if the oscillations produced by the microwave generator are counted, then when 9 192 631 770 oscillations have been detected, the time on the atomic clock is incremented by 1 s.

It now only remains to look at the details of the detection method used to measure the ratio of $|F_g = 4\rangle$ atoms to total atoms. The detection process generally utilizes the fluorescent signals produced by three horizontal laser beams tuned to different frequencies. The first beam that the atoms intercept is tuned to the energy of the $|F_g = 4\rangle \rightarrow |F_e = 5\rangle$ transition. The resulting fluorescence photons are detected by a photodiode, and the integrated number of photons detected during the atom pulse is proportional to the number of atoms in the $|F_g = 4\rangle$ ground state. The atoms are then removed by a transverse traveling-wave field, leaving only the $|F_g = 3\rangle$ ground-state atoms in the beam. The two remaining lasers are often superimposed on one another and are tuned to the $|F_g = 3\rangle \rightarrow |F_e = 4\rangle$ and $|F_g = 4\rangle \rightarrow |F_e = 5\rangle$ transition energies, respectively. The combined effect of these two laser beams is to first pump the $|F_g = 3\rangle$ atoms into the $|F_g = 4\rangle$ state; these $|F_g = 4\rangle$ atoms are then detected by observing the fluorescence photons resulting from the $|F_g = 4\rangle \rightarrow |F_e = 5\rangle$ irradiation using a second photodiode. The time-integrated signal from this second photodiode is proportional to the number of $|F_g = 3\rangle$ ground-state atoms in the original beam. The detector therefore yields a measure of the ratio of $|F_g = 3\rangle$ atoms to $|F_g = 4\rangle$ atoms.

The efficiency of the microwave cavity in inducing $|F_g = 3\rangle \rightarrow |F_g = 4\rangle$ transitions in the ground-state atoms depends on how close the microwave frequency is to the resonant frequency of the ground-state hyperfine transition, i.e. 9 192 631 770 Hz. Thus, the signal from the detector is fed back into the microwave generator to adjust the frequency to optimize the ratio of $|F_g = 3\rangle$ atoms to $|F_g = 4\rangle$ atoms and to provide an accurate time signal from the generator.

Atomic fountain clocks are in common use as time standards in many countries. These time standards include the National Institute of Standards and Technology (NIST) in Boulder, the Paris Observatory, the Physikalisch-Technische Bundesanstalt in Germany, and the National Physical Laboratory in the United Kingdom. Figure 9.4 shows the atomic fountain clock that currently serves as the time standard in the United States. Today's atomic fountain clocks have a relative uncertainty of the order of $\delta\nu/\nu \approx 10^{-16}$ (Heavner *et al* 2014). This translates to an uncertainty of about 1 s in 300 million years.

Another, potentially more accurate, method of constructing a clock using ultracold atoms is described in section 9.4.

9.3 Optical lattices

Optical lattices are a means of confining atoms in a periodic array by the use of standing light waves. They allow the study of physical properties that have previously only been available through the use of crystalline lattices and they also allow for greater control of lattice parameters. Optical lattices were first proposed by the Soviet physicist Vladilen Letokhov (1939–2009) in 1968 (Lethokov 1968). Experiments to study the properties of optical lattices began in the late 1980s; see the reviews by Jessen and Deutsch (1996) and Phillips (1998).

Standing waves with periodic minima with a spacing equal to half the laser wavelength can be created by the interference of two counterpropagating laser beams, as described in section 5.5. This arrangement of standing waves is referred to as an optical lattice, as it represents a periodic array formed by an oscillating potential analogous to the potential in a crystalline lattice. The simple example described here represents a one-dimensional optical lattice. However, two-dimensional or three-dimensional optical lattices can also be formed by utilizing additional orthogonally placed counterpropagating pairs of lasers. An example of a two-dimensional optical lattice is illustrated in figure 9.5. The potential as shown in the figure is generally referred to as an 'egg-crate potential' because of its shape. This is distinguished from the shape of the two-dimensional potential often used to describe crystallographic lattices, referred to as the 'muffin-tin potential,' which has potential wells with flat bottoms, rather than the rounded bottoms of the egg-crate potential.

Two important properties of the optical lattice can be controlled by the experimental conditions, namely the depth of the potential wells and the periodicity of the lattice. The depth of the potential wells is related to the intensity of the laser light. Most commonly, the intensity of the laser radiation input into the experiment is controlled by an acousto-optic modulator (sometimes referred to as a Bragg cell). Figure 9.6 shows a simple diagram of an acousto-optic modulator. Light, which is input into the cell, interferes with acoustic oscillations produced by a piezoelectric transducer (usually in the radio-frequency range). The light is diffracted by the sound waves, which act as an optical grating. The amount of light that is diffracted is a function of the intensity of the sound waves, and the relative light intensity

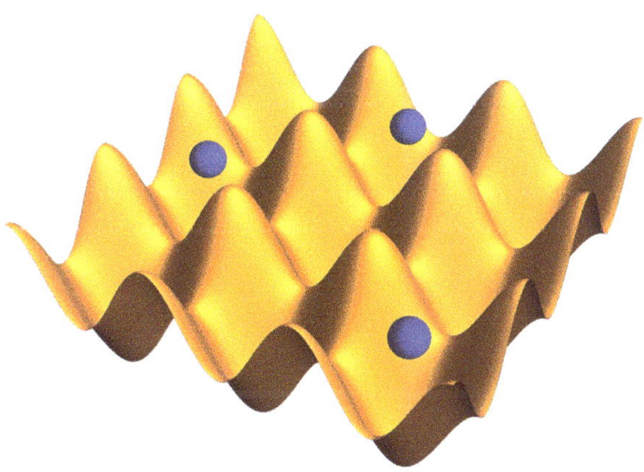

Figure 9.5. A two-dimensional optical lattice potential represented by the yellow surface, showing the equilibrial positions of trapped atoms represented by the blue spheres. This [2D–opticalAtoms (represented as blue spheres) pictured in a 2D-optical lattice potential (represented as the yellow surface)] image has been obtained by the author from the Wikimedia website where it was made available by Jpagett under a CC BY-SA 4.0 licence. It is included within this book on that basis. It is attributed to Jpagett.

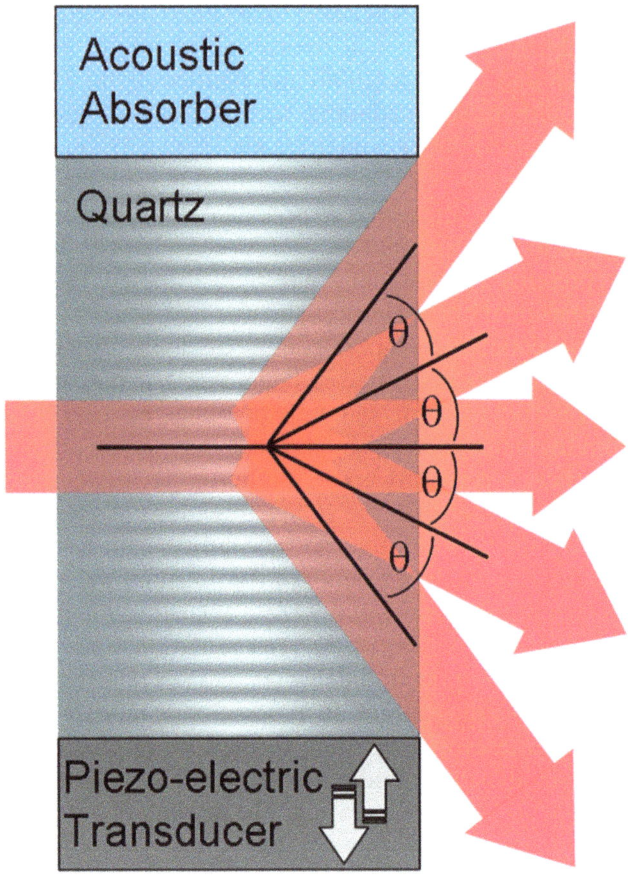

Figure 9.6. The principle of the acousto-optic modulator. This Cwbm (2009) image has been obtained by the author from the Wikimedia website, where it is stated to have been released into the public domain. It is included within this book on that basis.

produced by each order of diffraction can be varied by varying the intensity of the sound. Typically, for the zeroth-order beam, the intensity can be varied between about 15% and 99%, while for the first-order diffracted beam, the intensity can be varied between about 0% and 80%.

Control of the optical lattice spacing can be accomplished by two methods. Most obviously, it may be changed by adjusting the wavelength of the laser light. This may be done over the range of accessible wavelengths that can be provided by the laser. Ti:Al_2O_3 lasers, as discussed in section 5.4, are often used for this direct approach to controlling the optical lattice periodicity because of their large tunable range. Another fairly straightforward method of changing the periodicity of the optical lattice is to change the angle between the counterpropagating laser beams.

An important feature of an optical lattice is its ability to trap ultracold atoms. Cold atoms that are produced by the methods described in previous chapters can be introduced into the optical lattice, where they are trapped. The force on the atoms that results in their trapping within the optical lattice may be described in the following way. A shift in the atomic energy levels is caused by the Stark effect and results from the presence of the electric field associated with the standing wave. This shift is referred to as the light shift. The standing wave's light intensity, and hence electric field, is not homogeneous but has gradients, particularly near the minima and maxima. The resulting spatially dependent light shift gives rise to a dipole interaction that produces a force on the atom and traps it at a particular location relative to the optical lattice. If the laser is red detuned relative to the atomic transition energy, then the atom is trapped at a potential minimum in the optical lattice. If the laser is blue-detuned relative to the atomic transition energy, then the atom is trapped at a potential maximum in the optical lattice.

A couple of important points should be considered when conducting experiments using optical lattices. First, atoms which are trapped in potential wells associated with the lattice can hop from one site to another. This is a quantum process and occurs even when the atom's thermal energy is less than the depth of the potential well. The hopping probability can be reduced if the depth of the well is increased by increasing the laser intensity. Another factor to consider is the longevity of the distribution of atoms on the optical lattice. Once the cold atoms have been introduced into the optical lattice, they interact with the laser beams that were used to create the lattice and absorb energy. This raises the temperature of the atoms, which may be a factor that affects the duration available for experiments (see Grimm *et al* 2000). In the next section, some potential applications of optical lattices are described.

9.4 Optical lattice clocks

Atomic clocks are an interesting and important application of optical lattices (see e.g. Derevianko and Katori (2011), Poli *et al* (2013)). Although they are currently in the development stages, prototype devices have already surpassed the accuracy of the atomic fountain clocks described in section 9.2. The reason for the high accuracy of optical lattice clocks is that they utilize a transition in the optical region of the electromagnetic spectrum rather than one in the microwave region. This is readily seen by combining equations (9.1) and (9.2) for the Allan deviation, as follows:

$$\sigma_y \approx \frac{\Delta\nu}{\nu_0\sqrt{n\tau}}. \tag{9.4}$$

The frequency dependence in the denominator of equation (9.4) indicates that operating at higher frequencies provides a definite advantage in terms of the clock's accuracy. In fact, the difference between the microwave frequency used for ^{133}Cs clocks (9×10^9 Hz) and typical optical frequencies (4×10^{14} Hz) can amount to more than four orders of magnitude for the frequency in equation (9.4). In principle, this can translate into an uncertainty of 1 s over the age of the Universe.

A clock can be constructed that makes use of optical transitions in atoms that are in a gas (for example). However, as described in the previous section, trapping atoms in an optical lattice increases the coherent interaction time, Δt, of the atoms. Since the Fourier linewidth of the transition is related to the interaction time by

$$\Delta \nu \approx \frac{1}{\Delta t},$$

equation (9.4) shows that trapping the atoms improves clock accuracy. Other factors that contribute to the high accuracy of an optical lattice clock are the absence of a Doppler shift in the atomic energy levels and the ability to use larger samples (n in equation (9.4)) due to the minimal interactions between trapped atoms.

The clock transitions of optical lattice clocks can be electronic transitions in either neutral atoms or ions. Typical neutral atoms that are used for optical lattice clocks are those from groups 2A and 2B of the periodic table, or other atoms with two s-electrons outside the filled levels. Three of the most common atoms that have been utilized for optical lattice clocks are ^{87}Sr, ^{171}Yb, and ^{199}Hg. The properties of these atoms that are relevant to their use in optical lattice clocks are given in table 9.1.

The general design of an optical lattice clock is illustrated in figure 9.7 and an image of an experimental optical lattice clock is shown in figure 9.8. Atoms that

Table 9.1. Frequency standards recommended by the International Committee for Weights and Measures (Comité international des poids et mesures, CIPM) for three common neutral optical lattice clock atoms employing the $^1S_0 \rightarrow {}^3P_0$ transition. Data adapted from Riehle *et al* (2018), Poli *et al* (2013).

Atom	Electronic configuration	Clock frequency (Hz)	Fractional uncertainty	Cooling wavelength (nm)
^{87}Sr	[Kr]5s^2	429 228 004 229 873.0	4×10^{-16}	461 689
^{171}Yb	[Xe]4f^{14}5d°6s^2	518 295 836 590 863.6	5×10^{-16}	399 556
^{199}Hg	[Xe]4f^{14}5d^{10}6s^2	1 128 575 290 808 154.4	5×10^{-16}	253.7

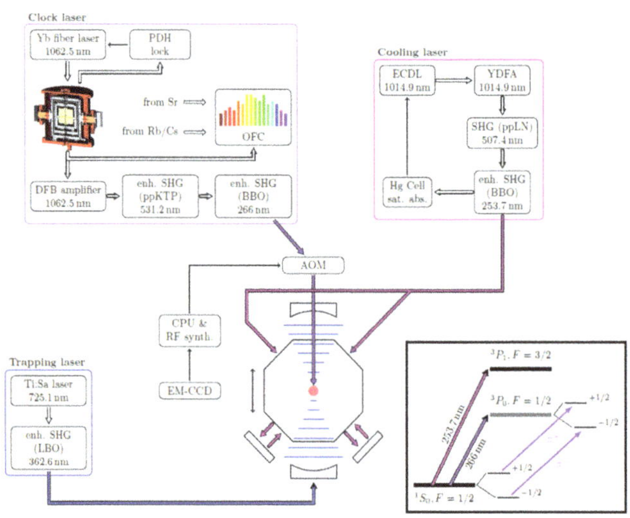

Figure 9.7. A typical design of an optical lattice clock, in this case operating on the $^1S_0 \rightarrow {}^3P_0$ transition of ^{199}Hg. AOM = acousto-optic modulator. Reprinted from De Sarlo *et al* (2016) CC BY 3.0. © IOP Publishing. Reproduced with permission. All rights reserved.

Figure 9.8. The ytterbium optical lattice clock at NIST. Laser light is input into the device through five optical fibers (such as the yellow optical fiber seen near the bottom center of the image). Reproduced from NIST (2013). Image stated to be in the public domain.

include an appropriate clock transition are cooled and trapped in an optical lattice as described in the previous section. As for atomic fountain clocks, a clock laser tuned to the appropriate frequency causes electronic transitions in the trapped atoms. Knowing the frequency of the clock transition (sometimes referred to as the 'magic frequency') allows time to be determined by counting clock laser oscillations.

At present, the second is defined in terms of an integral number of wavelengths of the hyperfine $6^2S_{1/2}$ ground-state transition $|F_g = 3\rangle \rightarrow |F_g = 4\rangle$ of ^{133}Cs at 9 192 631 770 Hz. For a typical optical transition (e.g. the $^1S_0 \rightarrow {}^3P_0$ transition in ^{87}Sr), the frequency is around 4×10^{14} Hz. The accuracy with which this frequency can be found determines the accuracy of the optical lattice clock that can be constructed. In the design described above, there are a number of challenges that need to be overcome in order to construct a functioning and useful device. Among these are the following:

How is the laser tuned to the proper clock frequency? And how are the clock laser oscillations accurately counted?

While these factors are also of relevance in the design of a microwave atomic fountain clock as described above, the approaches used to address these issues are not necessarily the same. Each of these two factors is discussed in some detail below.

9.4.1 The tuning of the clock laser

In order to understand the method used to tune the clock laser, we have to look in a little more detail at the energy level diagram of the atom. A typical situation, that of ^{87}Sr, is shown in figure 9.9. The diagram shows that, in addition to the clock

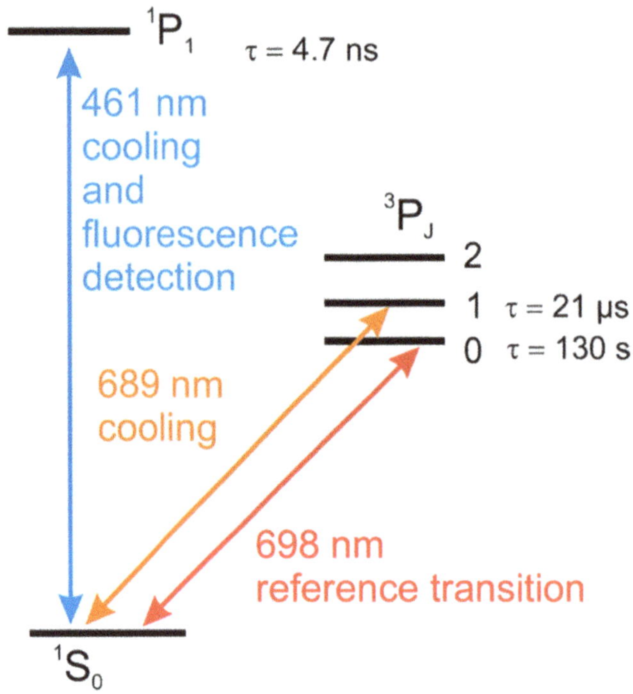

Figure 9.9. An energy level diagram of ^{87}Sr, showing the optical transitions of relevance to the construction of an optical lattice clock. The 698 nm reference transition is the clock transition. Reprinted from Riehle (2015), Copyright © 2015 Académie des sciences. Published by Elsevier Masson SAS. All rights reserved

transition and the cooling transition, a laser can also be utilized to induce fluorescence from the $^1S_0 \rightarrow {}^1P_1$ transition at 461 nm.

The fluorescence produced by the $^1S_0 \rightarrow {}^1P_1$ transition is observed with a CCD camera, as seen in figure 9.7. The fluorescent signal is a measure of the population of the $^1S_0$ ground state (as illustrated in figure 9.10) and is used as feedback for the frequency control of the clock laser. Figure 9.7 shows that the clock laser beam passes through an acousto-optic modulator (AOM, see figure 9.6) before entering the region of the optical lattice. The AOM, as described previously for intensity control, can also control the frequency of light. This is because the diffraction grating that is produced by the density fluctuations caused by the acoustic wave moves relative to the reference frame of the laser. This introduces a Doppler shift to the laser frequency given by

$$\Delta f = \pm m f_a,$$

where m is the order of the diffraction and f_a is the acoustic frequency. This is used as a means of adjusting the laser frequency to the resonant frequency of the $^1S_0 \rightarrow {}^1P_1$ transition and is analogous to the electronic feedback provided to the microwave generator in the atomic fountain clock described above.

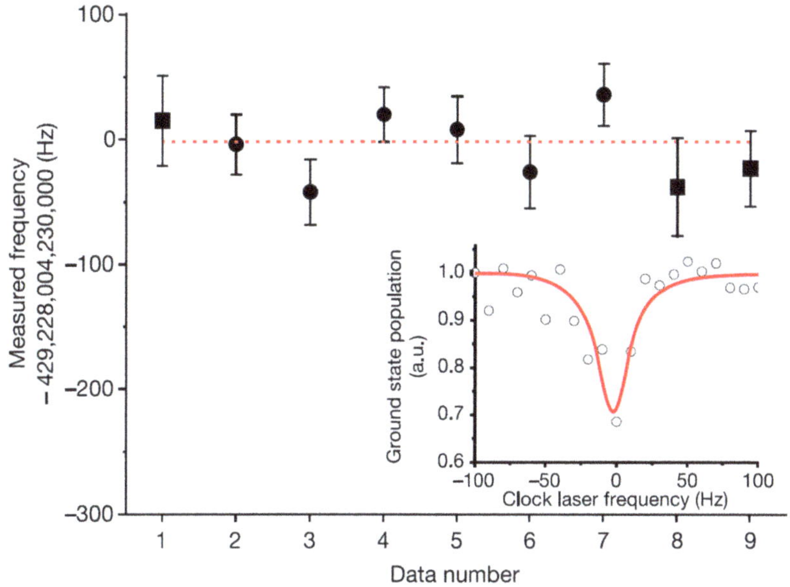

Figure 9.10. Scatter in the measured frequency of the $^1S_0 \rightarrow {}^3P_0$ transition of ^{87}Sr over a number of measurements. The inset shows the ground-state population as a function of the clock laser frequency and illustrates how the feedback signal can be used to tune the clock laser. Takamoto *et al* (2005) Copyright © 2005, Macmillan Magazines Ltd. Reproduced with permission from Springer Nature.

9.4.2 Counting the clock laser oscillations

The second point that is considered here is the method by which the oscillations of the clock laser can be counted. For microwave atomic clocks that operate in the 10^{10} Hz range, the microwave oscillations can be directly counted by an electronic counter. However, this is not possible at optical frequencies, which are in the range of 10^{14} Hz. It was not until the development of the optical frequency comb in the early 2000s that optical clocks became possible. We begin with a description of mode locking, which is essential for the creation of an optical frequency comb.

Mode locking is a technique by which a laser can be made to produce very short light pulses. From equation (1.3) and the relationship between frequency, f, and wavelength, λ, for an empty cavity,

$$\lambda = \frac{c}{f},$$

we can obtain the frequency of the modes as follows:

$$f = n\frac{c}{2L},$$

where n is an integer and L is the length of the cavity. This means that the frequency spacing between modes is

$$\Delta f = \frac{c}{2L}. \tag{9.5}$$

In a normal laser, each of the resonant cavity modes oscillates and creates a coherent and intense beam independently. Mode locking refers to a technique by which the phase relationship between the various modes is fixed. In this situation, the light produced by the laser is emitted in a train of very short pulses, each of which has a well-defined frequency. The length of the train is determined by the bandwidth of the lasing medium and the spacing (in frequency) between the pulses is given by equation (9.5). This situation is described graphically in figure 9.11.

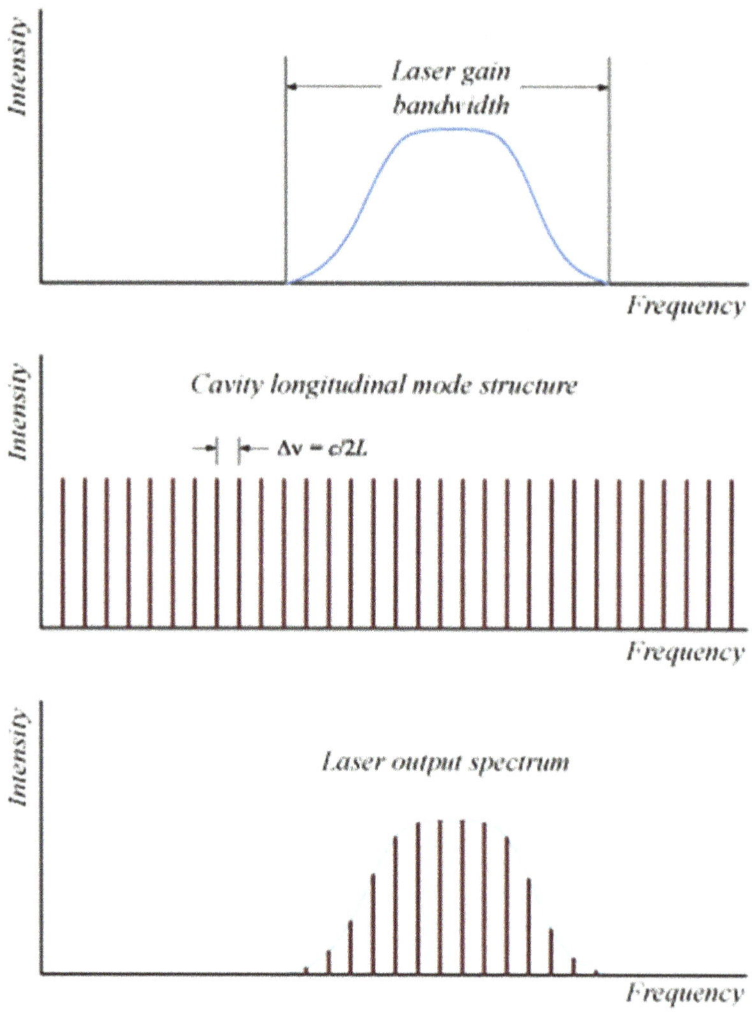

Figure 9.11. The properties of a mode-locked laser. (top) The bandwidth of the laser medium. (middle) The longitudinal cavity mode structure as a function of frequency. (bottom) The laser output pulse train. This DrBob (2003) image has been obtained by the author from the Wikimedia website where it was made available by Kelly under a CC BY-SA 3.0 licence. It is included within this book on that basis. It is attributed to DrBob.

A common method of mode locking a laser is by the use of an electro-optic modulator placed within the laser cavity. The electro-optic effect causes a change in the optical properties of a material in response to the application of an electric field at a frequency that is much lower than the frequency of light. In the case of the electro-optic modulator used for laser mode locking, the application of a sinusoidally varying electric field will induce an oscillating amplitude modulation of the light in the laser cavity. If the frequency of the laser light produced by a particular cavity mode is f and the modulation frequency is f_{rep}, then the light will have sidebands at the sum and difference frequencies, $f - f_{\text{rep}}$ and $f + f_{\text{rep}}$. If the modulation frequency is chosen to be

$$f_{\text{rep}} = \frac{c}{2L} \tag{9.6}$$

then the sidebands will correspond to the adjacent cavity modes, as defined in equation (9.5), and the central frequency and the sidebands will all be in phase. Driving the modulator at the sideband frequencies will generate additional in-phase sidebands at frequencies of $f - 2f_{\text{rep}}$ and $f + 2f_{\text{rep}}$. Repeating this approach for all the modes within the laser bandwidth will produce a laser output in which all the cavity modes are in phase. Equation (9.6) can also be viewed in the time domain, in which the inverse of the modulation frequency is defined to be the round-trip time of the laser pulse in the resonant cavity. This is discussed further below.

A mode-locked laser is the most common means of producing an optical frequency comb, as illustrated in figure 9.12. The name is derived from the comb-like appearance of the frequency spectrum. It can easily be seen from the figure that the frequency of the nth mode is given by

$$f_{\text{c}} = nf_{\text{rep}} + f_{\text{CEO}}, \tag{9.7}$$

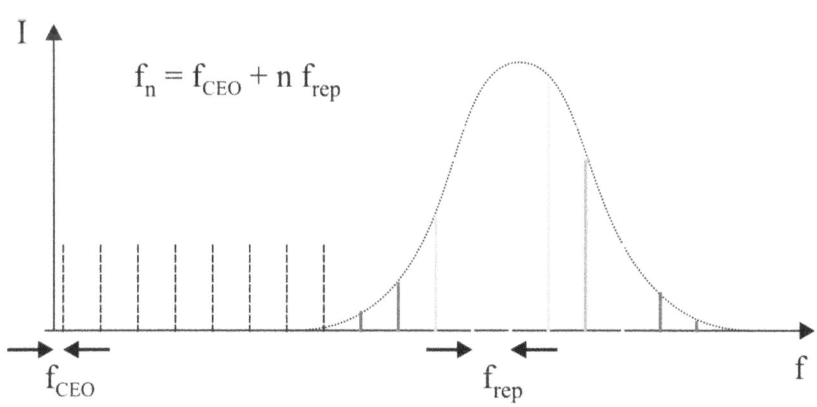

Figure 9.12. Optical frequency comb, showing the definition of f_{CEO} and f_{rep}. n is not to scale. This Hartmut (2006) image has been obtained by the author from the Wikimedia website where it was made available under a CC BY-SA 3.0 licence. It is included within this book on that basis. It is attributed to HartmutG.

where n is an integer (typically around 5×10^5) and f_{rep} is the frequency spacing of the peaks in the comb (referred to as the repetition rate), which is equal to the mode-locked laser's modulation frequency. f_{CEO} is the carrier-envelope offset frequency, which, by definition, is less than f_{rep}, and gives the frequency offset of the first mode.

The repetition frequency is measured against a Cs microwave clock standard as seen in figure 9.7. The repetition rate can be stabilized using a piezoelectric transducer, which adjusts the cavity length by moving one of the end mirrors. In this way, the oscillations of the repetition frequency can readily be counted and related to the clock frequency by equation (9.7).

9.5 The applications of optical lattice clocks

The development of optical clocks and, in particular, optical lattice clocks, has progressed substantially in recent years. This progress is shown in figure 9.13 in comparison to the progress in Cs microwave clocks based on measured or estimated fractional uncertainty. It can be seen that since around 2008, optical clocks have achieved a higher accuracy than microwave-frequency clocks. In the foreseeable future, it is expected that optical lattice clocks may achieve a fractional uncertainty of 10^{-19}, three orders of magnitude better than current Cs time standards. It is also likely that in the future, when optical lattice clocks achieve a suitable level of

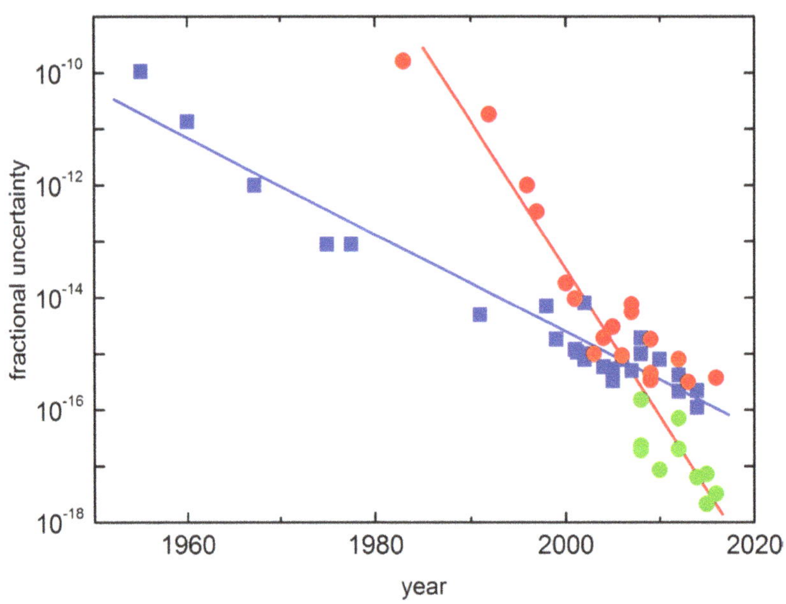

Figure 9.13. The fractional uncertainty of atomic clocks from the late 1950s to the late 2010s. Squares (blue) represent ^{133}Cs clocks. Circles represent clocks that utilize optical transitions (not only optical lattice clocks). Red circles show the results obtained by direct comparison to the Cs standard. Green circles represent estimated uncertainties. Reproduced from Riehle *et al* (2018). Copyright 2018 BIPM & IOP Publishing Ltd. All rights reserved.

reliability, they will replace Cs-based clocks as a means of providing the standard definition of the second.

In addition to providing a highly accurate time standard, optical lattice clocks have a number of other important applications. These include applications such as improved GPSs and geodesy. They also include studies of fundamental physical phenomena, such as special and general relativity. One of the interesting uses of high-accuracy optical lattices has been the investigation of the constancy of fundamental physical constants. Some cosmological theories predict that fundamental constants, such as the fine structure constant, vary with time (Uzan 2003) and atomic clocks have provided the most stringent tests to date for such variations. Research to date has considered two fundamental constants, the fine structure constant, α, and the electron-to-proton mass ratio, $\mu = m_e/m_p$. Some recent results are summarized below.

The energy scale for atomic transitions is defined by the Rydberg constant, R_∞, where

$$R_\infty = \frac{c}{4\pi\hbar}\alpha^2 m_e \tag{9.8}$$

and $\hbar$ is the reduced Planck constant. The fine structure constant, α, in equation (9.8) is defined as

$$\alpha = \frac{1}{4\pi\varepsilon_0}\frac{e^2}{\hbar c},$$

where ε_0 is the vacuum permittivity. The fine structure constant has a number of different but equivalent physical interpretations, but was originally viewed as the ratio of the velocity of the electron in the first circular orbit of the Bohr atom to the speed of light (Sommerfeld 1921).

Temporal variations in the fine structure constant would result in variations in the frequency of optical transitions as indicated by the definition of the Rydberg constant. These could be experimentally observed as changes in the ratio of the frequencies of two different transitions, R (Godun *et al* 2014). The time derivative of the fine structure constant, $d\alpha/dt$, is related to the time derivative of the frequency ratio, dR/dt as follows (Safronova *et al* 2018):

$$\frac{1}{R}\frac{dR}{dt} = (K_2 - K_1)\frac{1}{\alpha}\frac{d\alpha}{dt},$$

where K_1 and K_2 are the sensitivity coefficients of the two optical transitions.

Variations in the electron-to-proton mass ratio can be observed as changes in the frequency of hyperfine split ground-state levels, as utilized in Cs-based clocks, relative to a suitable optical transition frequency (Huntemann *et al* 2014). The time derivative of the electron-to-proton mass ratio, $d\mu/dt$, is expressed as (Godun *et al* 2014, Huntemann *et al* 2014, Safronova *et al* 2018)

$$\frac{1}{R}\frac{dR}{dt} \approx (K - K_{Cs})\frac{1}{\alpha}\frac{d\alpha}{dt} + \frac{1}{\mu}\frac{d\mu}{dt},$$

owhere K and K_{Cs} are the sensitivity coefficients of the optical and the Cs transitions, respectively, and the higher-order terms related to variations in quark masses are not included.

A recent study by Lange *et al* (2021) has looked at variations of both the fine structure constant and the electron-to-proton mass ratio. Figure 9.14 shows the results of an investigation of variations in the fine structure constant as evidenced by the ratio of frequencies for the $^2S_{1/2}(F = 0) \rightarrow {}^2D_{3/2}(F = 2)$ and $^2S_{1/}2(F = 0) \rightarrow {}^2F_{7/2}(F = 3)$ transitions in a $^{171}Yb^+$ ion. This experiment was conducted over a period of 1500 days. The horizontal gray band shows the statistical uncertainties over the duration of the experiment. The broken red line represents a linear fit to the data added in order to observe any temporal variations of the fine structure constant. The broken blue line is a fit to an annual sinusoidal oscillation representing coupling to the gravitational field. The annual variation in the coupling to gravity results from the eccentricity of the Earth's orbit. The results shown in figure 9.14 place limits on the fractional temporal variations of the fine structure constant of $1.0 \pm 1.1 \times 10^{-18}$ per year.

Figure 9.15 shows results from Lange *et al* (2021) for measurements of the frequency of the $^2S_{1/}2(F = 0) \rightarrow {}^2F_{7/2}(F = 3)$ transitions in a $^{171}Yb^+$ ion. The data are referenced to the recommended standard frequency for this transition of 642 121 496 772 645 Hz (Riehle *et al* 2018) based on a Cs fountain clock. The data presented in

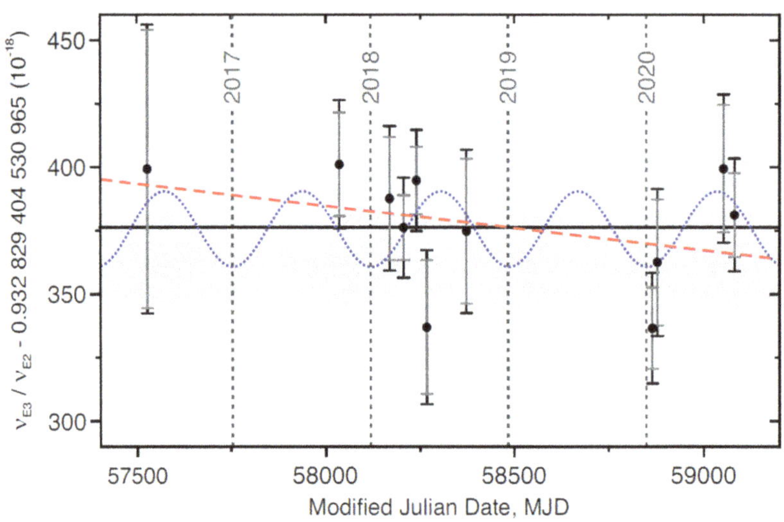

Figure 9.14. The measured ratio of frequencies, R, for the $^2S_{1/2}(F = 0) \rightarrow {}^2D_{3/2}(F = 2)$ transition to the $^2S_{1/}2$ $(F = 0) \rightarrow {}^2F_{7/2}(F = 3)$ transition in a $^{171}Yb^+$ ion. The full data set gives an average of R = 0.932 829 404 530 965 376(32), and the vertical axis of the figure shows a plot of the last three digits of the 18-digit measured ratio, where the solid black line shows the average value. The gray band shows the statistical uncertainty of the measurements. The broken red line is a fit to a linear temporal variation, and the broken blue line is a fit to an annual sinusoidal oscillation as described in the text. CC BY 4.0. Reprinted with permission from Lange *et al* (2021) Copyright (2021) by the American Physical Society.

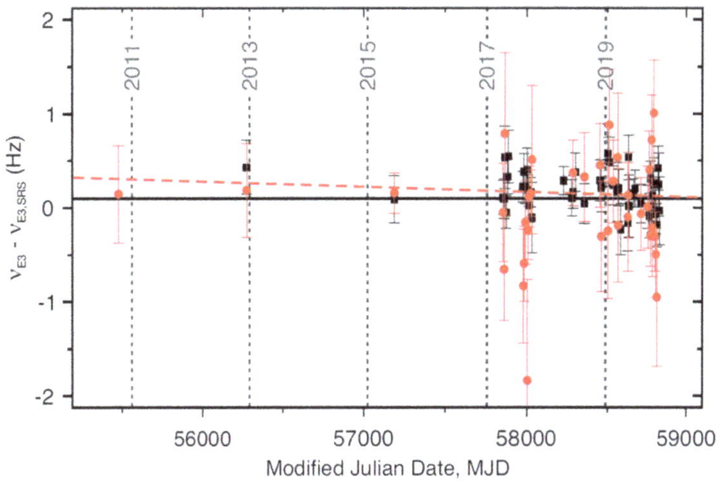

Figure 9.15. The measured frequency of the $^2S_{1/2}(F = 0) \rightarrow {}^2F_{7/2}(F = 3)$ transitions in a $^{171}Yb^+$ ion referenced to the frequency of a Cs fountain clock as described in the text. The measurements were carried out over a period of about 3350 days. The solid black line shows the average value of the full data set, and the broken red line shows a fit to a linear temporal variation. CC BY 4.0. Reprinted with permission from Lange *et al* (2021) Copyright (2021) by the American Physical Society.

the figure put limits on the fractional temporal variation of the electron-to-proton mass ratio of $-8 \pm 36 \times 10^{-18}$ per year.

The coupling of the fine structure constant and the electron-to-proton mass ratio to gravity can be established from the data presented in figures 9.14 and 9.15, respectively. In the case of the fine structure constant, the data in figure 9.14 have been fitted to a sinusoidal oscillation of the form

$$R = A \cos\left[2\pi\frac{(t - t_p)}{T}\right] + B,$$

where A and B are the fitted parameters, t_p is the time of the perihelion in the year 2018, and T is the length of the year. The results of this fit provide a limit on the possible coupling of the fine structure constant to gravity of

$$\frac{c^2}{\alpha}\frac{d\alpha}{d\Phi} = 1.4 \pm 1.1 \times 10^{-8},$$

where Φ is the Sun's gravitational potential.

For the electron-to-proton mass ratio, the data in figure 9.15 provide a limit on the coupling to gravity of

$$\frac{c^2}{\mu}\frac{d\mu}{d\Phi} = 7 \pm 45 \times 10^{-8}.$$

The measurements described above do not rule out the possibility that the fundamental constants vary in time, either linearly, as suggested by some cosmological theories, or sinusoidally as a result of coupling to a gravitational field. However, they do represent the most stringent test of these possibilities to date. Future improvements in optical lattice clocks will undoubtedly place stricter limits on the temporal variations of the fundamental constants or will show that such variations exist.

References and suggestions for further reading

Allan D W 1966 Statistics of atomic frequency standards *Proc. IEEE* **54** 221–30

Barontini G *et al* 2022 Measuring the stability of fundamental constants with a network of clocks *EPJ Quantum Technol.* **9** 12

Bauch A 2002 Caesium atomic clocks: function, performance and applications *Meas. Sci. Technol.* **14** 1159

Cwbm 2009 Acousto-optic Modulator https://commons.wikimedia.org/wiki/File:Acousto-optic_Modulator.png

De Sarlo L, Favier M, Tyumenev R and Bize S 2016 A mercury optical lattice clock at LNE-SYRTE *J. Phys. Conf. Ser.* **723** 012017

Derevianko A and Katori H 2011 Physics of optical lattice clocks *Rev. Mod. Phys.* **83** 331–47

DrBob 2003 Cavity mode diagram https://commons.wikimedia.org/wiki/File:Modelock-1.png

Gao Q *et al* 2018 Systematic evaluation of a ^{171}Yb optical clock by synchronous comparison between two lattice systems *Sci. Rep.* **8** 8022

Godun R M, Nisbet-Jones P B R, Jones J M, King S A, Johnson L A M, Margolis H S, Szymaniec K, Lea S N, Bongs K and Gill P 2014 Frequency ratio of two optical clock transitions in ^{171}Yb^{+} and constraints on the time variation of fundamental constants *Phys. Rev. Lett.* **113** 210801

Grimm R, Weidemüller M and Ovchinnikov Y B 2000 Optical dipole traps for neutral atoms *Adv. At. Mol. Opt. Phys.* **42** 95–170

Hartmut G 2006 A frequency comb, generated by a train of light pulses, and the definition of the carrier-envelope frequency https://commons.wikimedia.org/wiki/File:FrequencyComb-CEOphase.svg

Heavner T P, Donley E A, Levi F, Costanzo G, Parker T E, Shirley J H, Ashby N, Barlow S and Jefferts S R 2014 First accuracy evaluation of NIST-F2 *Metrologia* **51** 174–82

Huntemann N, Lipphardt B, Tamm C, Gerginov V, Weyers S and Peik E 2014 Improved limit on a temporal variation of m_p/m_e from comparisons of Yb^{+} and Cs atomic clocks *Phys. Rev. Lett.* **113** 210802

Itano W M, Bergquist J C, Bollinger J J, Gilligan J M, Heinzen D J, Moore F L, Raizen M G and Wineland D J 1993 Quantum projection noise: population fluctuations in two-level systems *Phys. Rev.* A **47** 3554–70

Jessen P S and Deutsch I H 1996 Optical lattices *Adv. Atom. Mol. Opt. Phys.* **37** 95–138

Joyet A 2003 Aspects métrologiques d'une fontaine continue à atoms froids *PhD Thesis* Université de Neuchâtel https://bib.rero.ch/global/documents/559109

Jpagett 2020 Atoms (represented as blue spheres) pictured in a 2D-optical lattice potential (represented as the yellow surface). https://commons.wikimedia.org/wiki/File:AtomsInLattice.png

Katori H 2011 Optical lattice clocks and quantum metrology *Nat. Photon.* **5** 203–10

Katori H, Ido T and Kuwata-Gonokami M 1999 Optimal design of dipole potentials for efficient loading of Sr atoms *J. Phys. Soc. Jpn.* **68** 2479–82

Kim H, Heo M S, Park C Y, Yu D H and Lee W K 2021 Absolute frequency measurement of the ^{171}Yb optical lattice clock at KRISS using TAI for over a year *Metrologia* **58** 055007

Lange R, Huntemann N, Rahm J M, Sanner C, Shao H, Lipphardt B, Tamm C, Weyers S and Peik E 2021 Improved limits for violations of local position invariance from atomic clock comparisons *Phys. Rev. Lett.* **126** 011102

Lethokov V S 1968 Narrowing of the Doppler width in a standing light wave *JETP Lett.* **7** 272 http://jetpletters.ru/ps/1685/article_25663.shtml

Metcalf H and van der Straten P 1999 *Laser Cooling and Trapping* (New York: Springer)

NIST 2013 NIST's ultra-stable ytterbium lattice atomic clock https://www.nist.gov/news-events/news/2013/08/nist-ytterbium-atomic-clocks-set-record-stability

Phillips W D 1998 Nobel lecture: laser cooling and trapping of neutral atoms *Rev. Mod. Phys.* **70** 721–41

Poli N, Oates C W, Gill P and Tino G M 2013 Optical atomic clocks *Riv. Nuovo Cimento.* **36** 555–624

Riehle F 2015 Towards a redefinition of the second based on optical atomic clocks *C. R. Phys.* **16** 506–15

Riehle F, Gill P, Arias F and Robertsson L 2018 The CIPM list of recommended frequency standard values: guidelines and procedures *Metrologia* **55** 188–200

Rosenband T *et al* 2008 Frequency ratio of Al^+ and Hg^+ single-ion optical clocks; metrology at the 17th decimal place *Science* **319** 1808–12

Safronova M S, Budker D, DeMille D, Kimball D F J, Derevianko A and Clark C W 2018 Search for new physics with atoms and molecules *Rev. Mod. Phys.* **90** 025008

Sommerfeld A 1921 *Atombau und Spektrallinien* 2nd edn (Braunschweig: Druck und Verlag von Friedr. Vieweg & Sohn), 241–2

Stwalley W C 2004 Collisions and reactions of ultracold molecules *Can. J. Chem.* **82** 709–12

Takamoto M, Hong F L, Higashi R and Katori H 2005 An optical lattice clock *Nature* **435** 321–4

Taylor B N 2001 *The International System of Units (SI)* (Gaithersburg, MD: NIST Special Publication) https://physics.nist.gov/cuu/pdf/sp330.pdf NIST Special Publication 330

Udem T H, Reichert J, Holzwarth R and Hänsch T W 1999 Absolute optical frequency measurement of the cesium D1 line with a mode-locked laser *Phys. Rev. Lett.* **82** 3568–71

Uzan J P 2003 The fundamental constants and their variation: observational and theoretical status *Rev. Mod. Phys.* **75** 403–55

Weiner J, Bagnato V S, Zilio S and Julienne P S 1999 Experiments and theory in cold and ultracold collisions *Rev. Mod. Phys.* **71** 1–85

Wynands R and Weyers S 2005 Atomic fountain clocks *Metrologia* **42** S64

www.ingramcontent.com/pod-product-compliance
Ingram Content Group UK Ltd.
Pitfield, Milton Keynes, MK11 3LW, UK
UKHW051307130926
8556IPUK00048B/561

9 780750 354806